BE AGILE®

Scrum Training for Teams

Part of the Agile Education Series™

Agile Ice Breaker

The
Marshmallow
Challenge

20 sticks of spaghetti + one yard tape + one yard string + one marshmallow

3

CAPE PROJECT
MANAGEMENT, INC.

Notes:

Introductions and Expectations

CAPE PROJECT
MANAGEMENT, INC.

Notes:

About Us

- ‣ The course curriculum developed by Dan Tousignant, PMI-ACP, CSP of Cape Project Management, Inc.
- ‣ We provide public, onsite and online training:
  - ◦ AgileProjectManagementTraining.com
- ‣ Follow us on Twitter @ScrumDan
- ‣ The content of this course is licensed to your instructor.

5

Copyrighted material. 2018

CAPE PROJECT
MANAGEMENT, INC.

Notes:

The Agile Education Series™

1. Scrum Master Certification Training
2. Product Owner and User Story Training
3. All About Agile™: PMI–ACP® Agile Exam Preparation
4. Kanban for Software Development Teams
5. Achieving Agility – How to implement Agile in your organization
6. Agile for Team Members
7. Agile for Executives
8. Effective Agile Testing

▸ All of these curriculums are available on Amazon at:
  ◦ http://bit.ly/DansAgileBooks

CAPE PROJECT
MANAGEMENT, INC.

Notes:

Continuing Education

▸ This course provides education credits for the following certifications:
  ○ Continuing Certification for PMPs & PMI-ACPs: 7 PDUs Category B: Continuing Education
  ○ PMI-ACP Application: 7 Agile Education Contact Hours
  ○ Scrum Alliance SEUs for CSP Application and Maintenance: 7 SEUS Category C: Outside Events

7

CAPE PROJECT
MANAGEMENT, INC.

Notes:

Course Objectives

‣ Provide an in-depth review of the Scrum Framework

‣ Learn what it means to be on a Scrum Team

‣ Have fun!

8   CAPE PROJECT
MANAGEMENT, INC.

Notes:

Agenda

Modules
Why Agile?
The Agile Manifesto
The Scrum Framework
The Sprint Game
The Agile Product Lifecycle
Implementing Agile

9

CAPE PROJECT
MANAGEMENT, INC.

Notes:

Announcements

- Participant materials
  - Slides
  - Exercises
  - Glossary
- Facilities orientation
- Breaks

CAPE PROJECT
MANAGEMENT, INC.

Notes:

Typical Project Risks

▸ Misunderstanding of the requirements
▸ Lack of management commitment and support
▸ Lack of adequate user involvement
▸ Failure to gain user commitment
▸ Failure to manage end user expectation
▸ Changes to requirements
▸ Lack of an effective project management methodology

http://mooc.ee/MTAT.03.243/2015_spring/uploads/Main/top-10.pdf

11

CAPE PROJECT
MANAGEMENT, INC.

Notes:

Project Success Rates

	Successful	Challenged	Failed
Waterfall	11%	60%	29%
Agile	39%	52%	9%

■ Successful Challenged ■ Failed

Success= On-time, On-budget with a satisfactory result
(% of requirements met strategic goal)

https://www.infoq.com/articles/standish-chaos-2015

12

CAPE PROJECT
MANAGEMENT, INC.

Notes:

Why Agile is so Successful?

- ▸ Allows for more flexibility in requirements and development
- ▸ Includes built-in increments and iterations
- ▸ Provides the opportunity to encounter and address errors sooner in the development cycle
- ▸ Increases organizational and team efficiency
- ▸ Decreases unnecessary documentation and meetings
- ▸ Provides a value-based approach to development
- ▸ Assumes organizational differences
  - ◦ "Can be right sized"

13

CAPE PROJECT
MANAGEMENT, INC.

Notes:

Requirements Stability vs. Development Approach

- Selecting a development approach depends upon the stability of the requirements:
  - A predictive team can report exactly what features and tasks are planned for the entire length of the development process.
  - Adaptive methods focus on adapting quickly to changing realities. When the needs of a project change, an adaptive team changes as well.

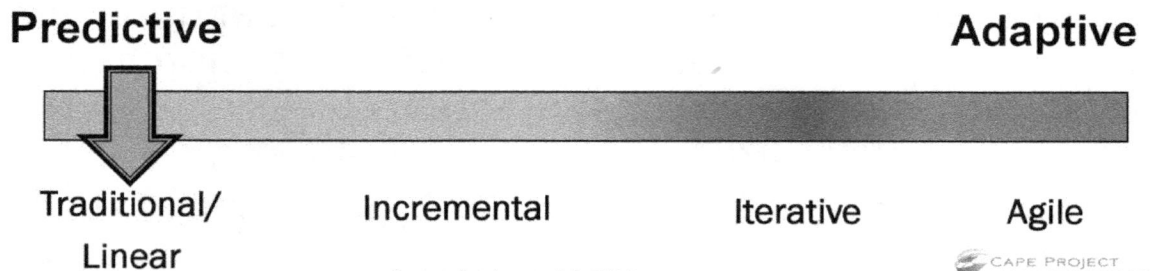

Predictive   **Adaptive**

Traditional/   Incremental   Iterative   Agile
Linear

14

Copyrighted material. 2018

CAPE PROJECT
MANAGEMENT, INC.

Notes:

DISCUSSION

Why be more Agile or Adaptive in your organization?

What types of requirement changes do you experience?

 CAPE PROJECT MANAGEMENT, INC.

Notes:

The Agile Manifesto

16

CAPE PROJECT
MANAGEMENT, INC.

CAPE PROJECT MANAGEMENT, INC.

Notes:

The Agile Manifesto: A statement of values

We are uncovering better ways of developing software by doing it and helping others do it.

Through this work we have come to value:

Individuals and interactions	over	Process and tools
Working software	over	Comprehensive documentation
Customer collaboration	over	Contract negotiation
Responding to change	over	Following a plan

That is, while there is value in the items on the right, we value the items on the left more.

Source: www.agilemanifesto.org

18

Copyrighted materials. 2018

CAPE PROJECT MANAGEMENT, INC.

Notes:

12 Principles of the Agile Manifesto

1. Our highest priority is to satisfy the customer through early and continuous delivery of valuable software.

2. Welcome changing requirements, even late in development. Agile processes harness change for the customer's competitive advantage.

3. Deliver working software frequently, from a couple of weeks to a couple of months, with a preference to the shorter timescale.

4. Business people and developers must work together daily throughout the project.

5. Build projects around motivated individuals. Give them the environment and support they need, and trust them to get the job done.

6. The most efficient and effective method of conveying information to and within a development team is face-to-face conversation.

7. Working software is the primary measure of progress.

8. Agile processes promote sustainable development. The sponsors, developers, and users should be able to maintain a constant pace indefinitely.

9. Continuous attention to technical excellence and good design enhances agility.

10. Simplicity—the art of maximizing the amount of work not done—is essential.

11. The best architectures, requirements, and designs emerge from self-organizing teams.

12. At regular intervals, the team reflects on how to become more effective, then tunes and adjusts its behavior accordingly.

CAPE PROJECT
MANAGEMENT, INC.

Notes:

Activity

Your 3 Principles

 CAPE PROJECT MANAGEMENT, INC.

Notes:

Activity: Agile Manifesto Principles

Directions:
1. Pair-up with another person at your table and review the Agile Manifesto Principles.
2. Pick three principles that you think are critical to the success of your Agile Implementation or are the most challenging.
3. Be prepared to share your answer with the class.

Principle	Choose 3
1. Our highest priority is to satisfy the customer through early and continuous delivery of valuable software.	
2. Welcome changing requirements, even late in development. Agile processes harness change for the customer's competitive advantage.	
3. Deliver working software frequently, from a couple of weeks to a couple of months, with a preference to the shorter timescale.	
4. Business people and developers must work together daily throughout the project.	
5. Build projects around motivated individuals. Give them the environment and support they need, and trust them to get the job done.	
6. The most efficient and effective method of conveying information to and within a development team is face-to-face conversation.	
7. Working software is the primary measure of progress.	

8. Agile processes promote sustainable development. The sponsors, developers, and users should be able to maintain a constant pace indefinitely.	
9. Continuous attention to technical excellence and good design enhances agility.	
10. Simplicity--the art of maximizing the amount of work not done--is essential.	
11. The best architectures, requirements, and designs emerge from self-organizing teams.	
12. At regular intervals, the team reflects on how to become more effective, then tunes and adjusts its behavior accordingly.	

Notes:

_____

_____

_____

_____

_____

_____

The Scrum Framework

23

CAPE PROJECT
MANAGEMENT, INC.

Scrum

- Scrum (n): A framework within which people can address complex adaptive problems, while productively and creatively delivering products of the highest possible value.

Source: Schwaber, Sutherland: *A Scrum Guide*

24

CAPE PROJECT
MANAGEMENT, INC.

Notes:

Origins of the idea

"The... 'relay race' approach to product development...may conflict with the goals of maximum speed and flexibility. Instead a holistic or 'rugby' approach—where a team tries to go the distance as a unit, passing the ball back and forth—may better serve today's competitive requirements."

Hirotaka Takeuchi and Ikujiro Nonaka, "The New New Product Development Game", *Harvard Business Review*, January 1986.

25

CAPE PROJECT MANAGEMENT, INC.

Notes:

History of Scrum

- Ken Schwaber and Jeff Sutherland developed the Scrum method in the early 1990's.
- In 2002 "Agile Software Development with Scrum" was written by Ken Schwaber & Mike Beedle
- The Scrum method has evolved somewhat over the years.
- 2002 – Present, organic growth due to its anecdotal successes, grass roots adoption and an aggressive training approach have made it the most common Agile methodology
- The definitive guide to the rules of Scrum, *The Scrum Guide*, is maintained by Ken Schwaber and Jeff Sutherland. http://www.Scrumguides.org

26

CAPE PROJECT
MANAGEMENT, INC.

Notes:

Scrum is:

- Scrum is a lightweight, simple to understand (but difficult to master) agile process framework.
- Scrum is one of several agile software development methods.
- Scrum and Extreme Programming (XP) are probably the two best-known Agile methods. XP emphasizes technical practices such as pair programming and continuous integration. Scrum emphasizes management practice such as the role of Scrum Master.
- Many companies use the management practices of Scrum with the technical practices of XP.

CAPE PROJECT
MANAGEMENT, INC.

Notes:

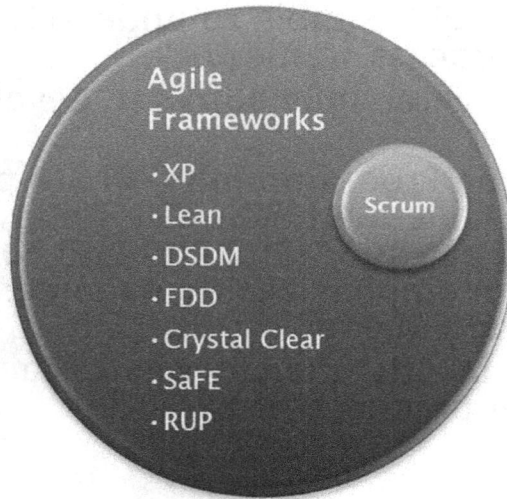

Scrum is Agile

Agile Frameworks

- XP
- Lean
- DSDM
- FDD
- Crystal Clear
- SaFE
- RUP

Scrum

CAPE PROJECT MANAGEMENT, INC.

Notes:

Scrum Adoption

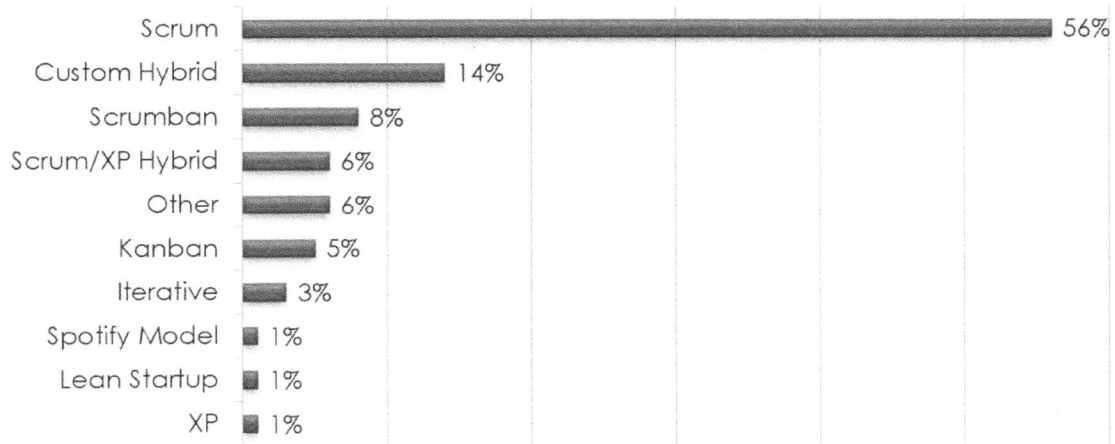

Method	Percentage
Scrum	56%
Custom Hybrid	14%
Scrumban	8%
Scrum/XP Hybrid	6%
Other	6%
Kanban	5%
Iterative	3%
Spotify Model	1%
Lean Startup	1%
XP	1%

Source: Version One 13th Annual Agile Survey, 2018

CAPE PROJECT MANAGEMENT, INC.

29

Copyrighted material. 2018

Notes:

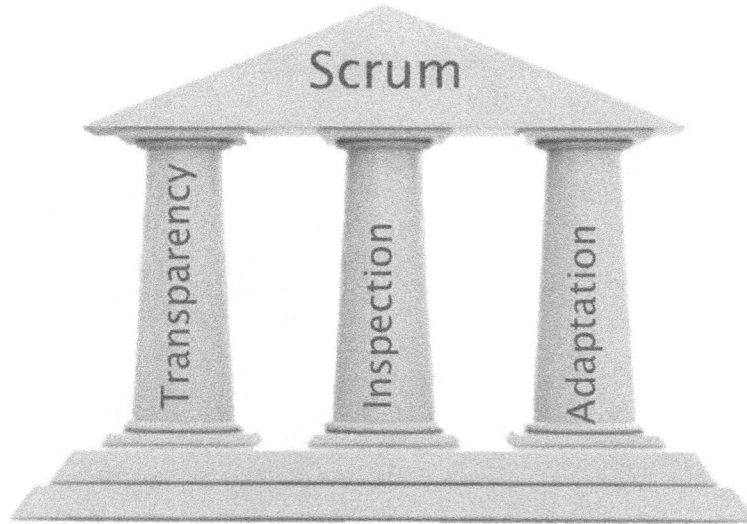

3 Pillars of Scrum

30

Notes:

3 Pillars of Scrum

Transparency
‣ ALL relative aspects of the process must be visible to those responsible for the outcome.

Inspection
‣ There is frequent inspection of the artifacts and progress to identify and correct undesirable variances. Inspection occurs during the Sprint Planning Meeting, Daily Scrum, Sprint Review and Sprint Retrospective.

Adaptation
‣ After inspection, adjustments should be made to the processes and artifacts to minimize further deviation.

31

CAPE PROJECT
MANAGEMENT, INC.

Notes:

5 Core Team Values of Scrum

1. **Commitment** – When we, as a team, value the commitment we make to ourselves and our teammates, we are much more likely to give our all to meet our goals.

2. **Focus** – When we value Focus, and devote the whole of our attention to only a few things at once, we deliver a better quality product, faster.

3. **Openness** – When we value being Open with ourselves and our teammates, we feel comfortable inspecting our behavior and practices, we can adapt them accordingly.

4. **Respect** – When we value Respect, people feel safe to voice concerns and discuss issues, knowing that their voices are heard and valued.

5. **Courage** – When we value Courage, people are encouraged to step outside of their comfort zones and take on greater challenges, knowing they will not be punished if they fail.

32

CAPE PROJECT MANAGEMENT, INC.

Notes:

Scrum Project Management

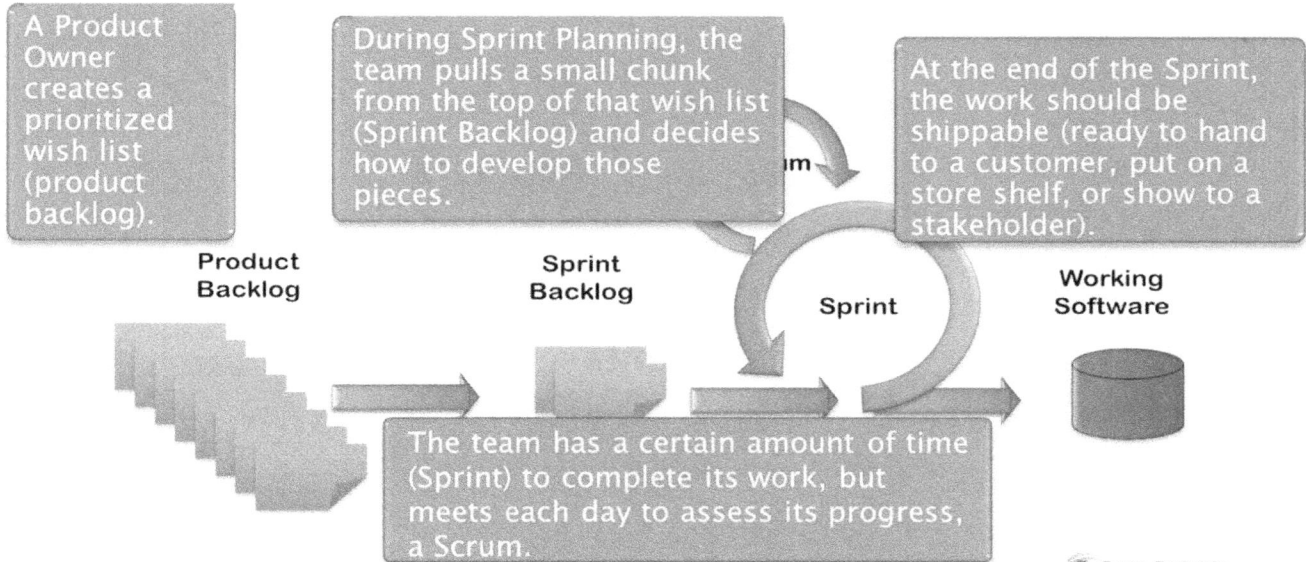

A Product Owner creates a prioritized wish list (product backlog).

During Sprint Planning, the team pulls a small chunk from the top of that wish list (Sprint Backlog) and decides how to develop those pieces.

At the end of the Sprint, the work should be shippable (ready to hand to a customer, put on a store shelf, or show to a stakeholder).

Product Backlog

Sprint Backlog

Sprint

Working Software

The team has a certain amount of time (Sprint) to complete its work, but meets each day to assess its progress, a Scrum.

CAPE PROJECT MANAGEMENT, INC.

33

Notes:

The Scrum Framework

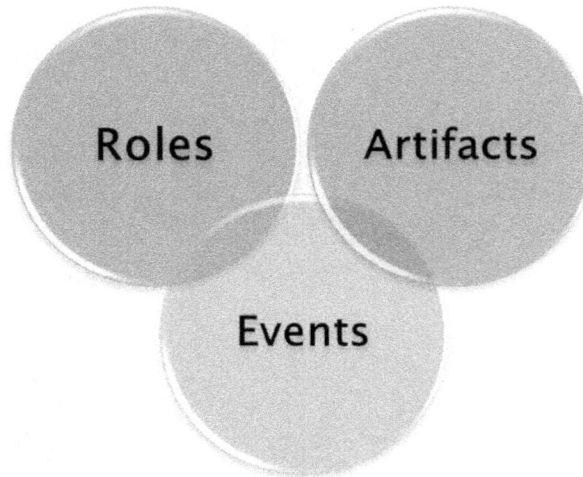

Roles

Artifacts

Events

CAPE PROJECT
MANAGEMENT, INC.

Notes:

The Scrum Framework

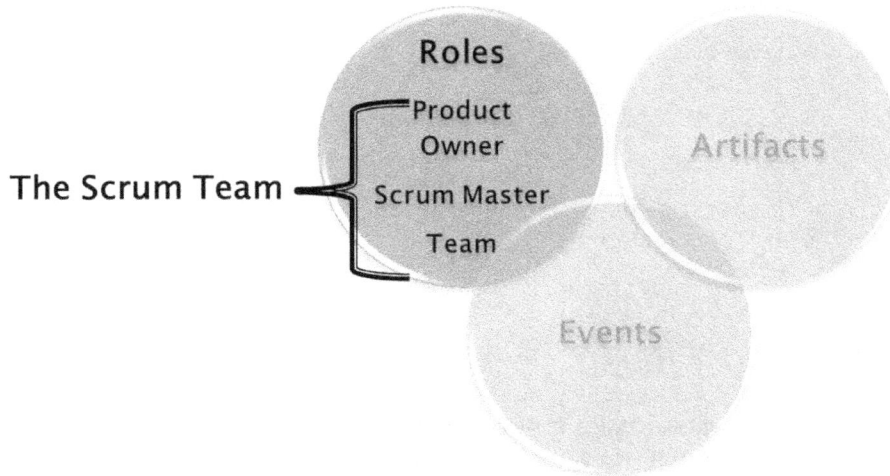

The Scrum Team
- Roles
  - Product Owner
  - Scrum Master
  - Team
- Artifacts
- Events

CAPE PROJECT
MANAGEMENT. INC.

Notes:

Scrum Master Responsibilities

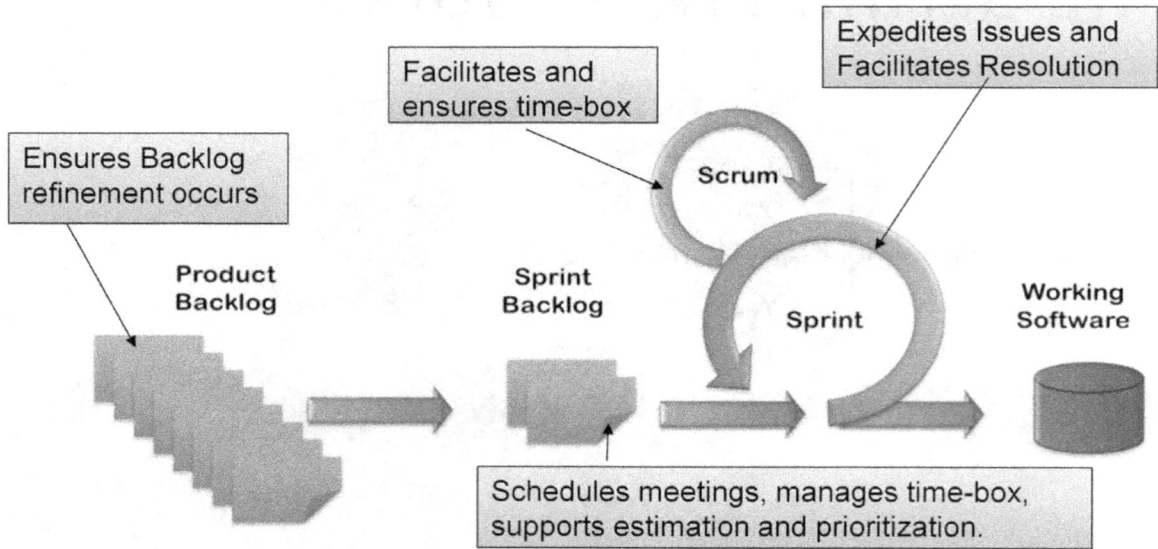

Ensures Backlog refinement occurs

Facilitates and ensures time-box

Expedites Issues and Facilitates Resolution

Scrum

Product Backlog

Sprint Backlog

Sprint

Working Software

Schedules meetings, manages time-box, supports estimation and prioritization.

36

Copyrighted material. 2018

CAPE PROJECT MANAGEMENT, INC.

Notes:

The Scrum Master Dos

- Represents management to the project
- Responsible for enacting Scrum values and practices
- Removes impediments
- Ensure that the team is fully functional and productive
- Enable close cooperation across all roles and functions
- Shield the team from external interferences
- Is a Servant Leader

CAPE PROJECT MANAGEMENT, INC.

Notes:

Servant Leadership Skills

▸ Traditional leadership is "command-and-control"
  ◦ "Workers need to be monitored closely"
▸ Servant leadership is based upon trust
  ◦ "Team members are self-motivated"
▸ An agile servant leader needs to:
  ◦ Protect the team from outside distractions
  ◦ Remove impediments to the team's performance
  ◦ Communicate and re-communicate project vision
  ◦ "Move boulders and carry water"—in other words, remove obstacles that prevent the team from providing business value

38

CAPE PROJECT
MANAGEMENT, INC.

Notes:

Scrum Master Don'ts

- Own the product decisions on Product Owner's behalf
- Make estimates on team's behalf
- Make the technology decisions on team's behalf
- Assign the tasks to the team members
- Try to manage the team

CAPE PROJECT
MANAGEMENT, INC.

Notes:

Scrum Master – Top 5

1. Coach team members
2. Manage conflict
3. Facilitate decision making
4. Remove team impediments
5. Increase organizational awareness of Scrum

CAPE PROJECT
MANAGEMENT, INC.

Notes:

Product Owner Responsibilities

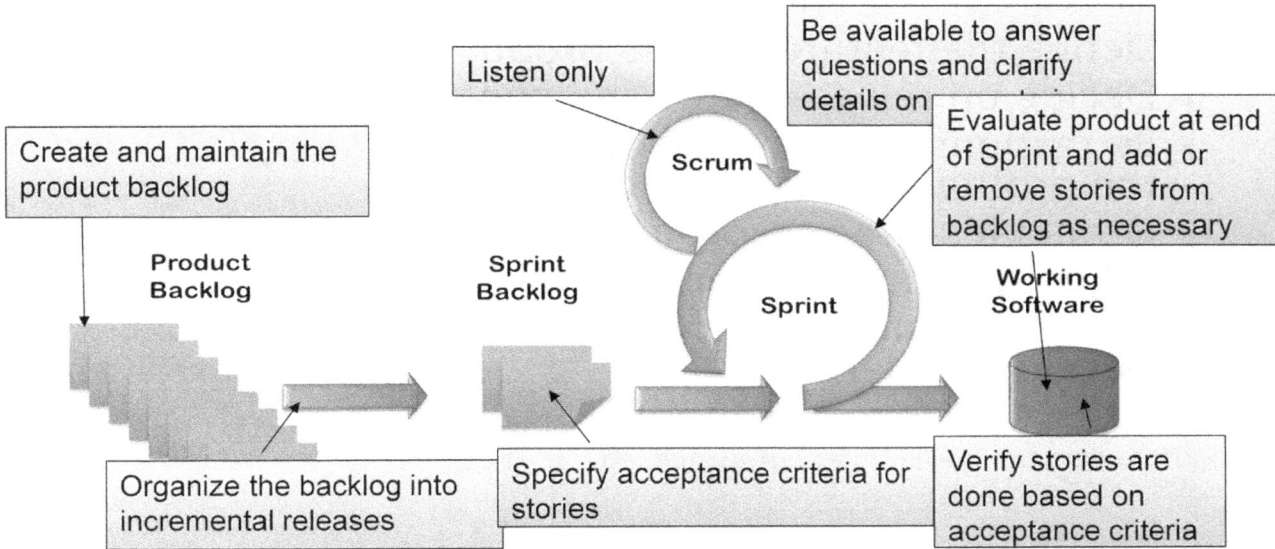

Listen only

Be available to answer questions and clarify details on

Create and maintain the product backlog

Evaluate product at end of Sprint and add or remove stories from backlog as necessary

Scrum

Product Backlog

Sprint Backlog

Sprint

Working Software

Organize the backlog into incremental releases

Specify acceptance criteria for stories

Verify stories are done based on acceptance criteria

41

Copyrighted material. 2018

CAPE PROJECT MANAGEMENT, INC.

Notes:

Product Owner Dos

- Define the features of the product
- Decide on release dates and content
- Be responsible the total cost of ownership (TCO) and the profitability of the product (ROI)
- Orders features according to market value
- Adjust features and priority every Sprint, as needed
- Available daily* to answer questions for the Development Team

*not necessarily full-time, but as needed

42   Copyrighted material. 2018   CAPE PROJECT MANAGEMENT, INC.

Notes:

Product Owner Don'ts

- Choose how much work will be accomplished in the Sprint – the team will do this, based on the priorities
- Change anything within the Sprint once it has started and don't add items unless the Sprint will end early
- Answer the three questions at the Daily Scrum Meetings (unless they have a task on a User Story)

43

CAPE PROJECT
MANAGEMENT, INC.

Notes:

Product Owner – Top 5

1. Maximizes the value of the product backlog
2. Represent the customer in Scrum events
3. Ensure the team is adapting to change
4. Inspect product progress
5. Accept or reject work results

CAPE PROJECT
MANAGEMENT, INC.

Notes:

Product Owner vs Scrum Master

Two different roles that complement each other. If one is not played properly, the other suffers.

- The Product Owner is responsible for the *product success*
- The Scrum Master is responsible for *project success*

CAPE PROJECT
MANAGEMENT, INC.

Notes:

© Scott Adams, Inc./Dist. by UFS, Inc.

CAPE PROJECT MANAGEMENT, INC.

Notes:

Team Responsibilities

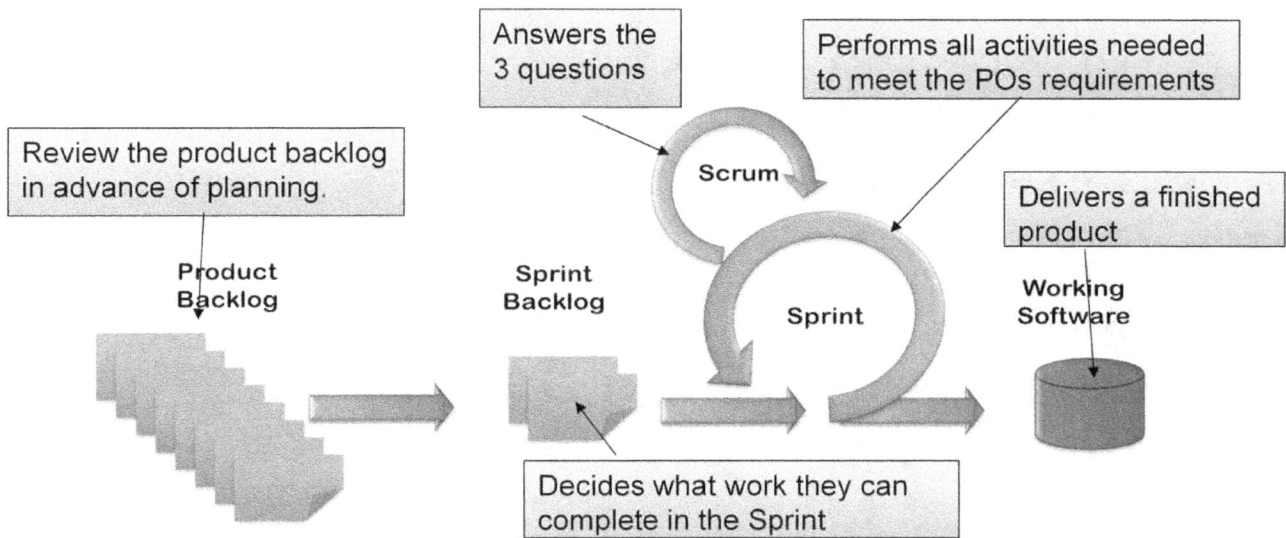

Answers the 3 questions

Performs all activities needed to meet the POs requirements

Review the product backlog in advance of planning.

Delivers a finished product

Product Backlog

Sprint Backlog

Scrum

Sprint

Working Software

Decides what work they can complete in the Sprint

47

Copyrighted material. 2018

CAPE PROJECT MANAGEMENT, INC.

Notes:

The Team Dos

- Typically 6 ± 3 people
- Teams are self-organizing
- Cross-functional:
  - Programmers, testers, user experience designers, etc.
- Members should be full-time
  - May be exceptions (e.g., database administrator)
- Membership should change only between Sprints

48

CAPE PROJECT
MANAGEMENT, INC.

Notes:

The Team Don'ts

- Skip Scrum Meetings
- Stop working when there is a roadblock or not enough information
- Increase technical debt in order to meet the velocity
- Individuals on a Scrum team should not do excessive individual overtime, or in any other way try to be the "hero" of the team.
  - Scrum helps us build *great teams of people, not teams of great people*
  - ~ Barry Turner

49

CAPE PROJECT MANAGEMENT, INC.

Notes:

What is a self-organizing team?

- Team members are self-empowered (they know what needs to be done and have the means to do it)
- They are willing to take on the responsibility of self-organizing and self-examining
- They are ready to let go of their individual egos and differences so that the team can function together
- Scrum's success is dependent on the team's ability to self-examine itself and to continually seek improvement

http://blog.openviewpartners.com/scrum-challenge-self-organizing-teams/

50

CAPE PROJECT
MANAGEMENT, INC.

Notes:

Self-organizing Team Challenges

- ▸ One of the hardest challenge of implementing Scrum
- ▸ Typically the biggest cultural change
- ▸ Full of "Scrum-Buts"
- ▸ Existing problems are often highlighted or exacerbated

CAPE PROJECT
MANAGEMENT, INC.

Notes:

Traditional versus Agile Teams

Traditional Teams

Project Manager →

Agile Teams

Self Organizing

Servant Leader →

52

Copyrighted material. 2018

CAPE PROJECT MANAGEMENT, INC.

Notes:

The Team Top 5

1. Self-organizing team members have to be more creative
2. Have to have a strong discipline and work ethic
3. Must be committed to the project's goals
4. Respect each other
5. Share a genuine conviction that the "we"—the potent concept behind every team—will succeed or fail together

CAPE PROJECT
MANAGEMENT, INC.

Notes:

Benefits of a Self-organizing Team

- ‣ Promotes creativity and problem solving throughout the organization
- ‣ Speeds adaptation to change
- ‣ Generates high-quality products and services
- ‣ Decreases the odds of burnout
- ‣ Supports leadership among peers

54   Copyrighted material. 2018   CAPE PROJECT MANAGEMENT, INC.

Notes:

How Teams Develop

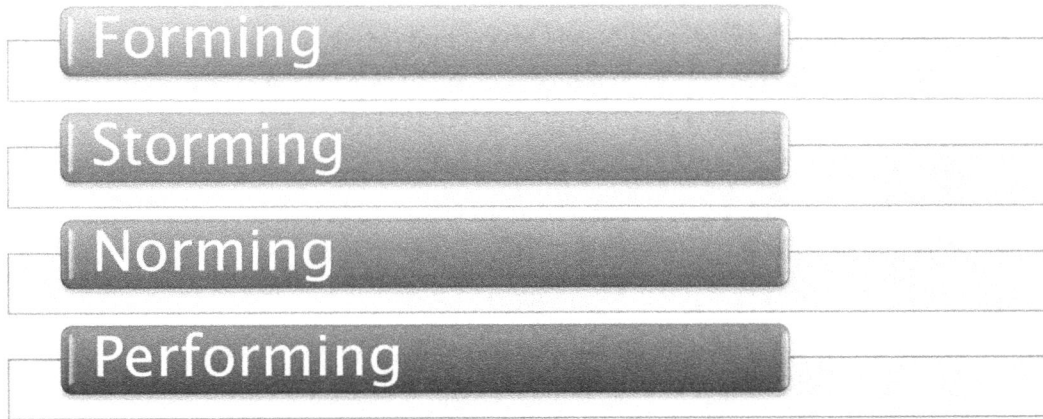

- Forming
- Storming
- Norming
- Performing

55

CAPE PROJECT MANAGEMENT, INC.

Notes:

Forming

‣ In this stage, most team members are positive and polite. Some are anxious, as they haven't fully understood what work the team will do. Others are simply excited about the task ahead.

‣ The leader will play a dominant role at this stage, because team members' roles and responsibilities aren't clear.

‣ This stage can last for some time, as people start to work together, and as they make an effort to get to know their new colleagues.

Copyrighted material. 2018

CAPE PROJECT
MANAGEMENT, INC.

Notes:

Storming

- ‣ This is the stage where many teams fail.
- ‣ Storming often starts where there is a conflict between team members' natural working styles.
- ‣ If it has not been clearly defined how the team will work, people may feel overwhelmed by their workload.
- ‣ Some people may question the worth of the team's goal, and they may resist taking on tasks.
- ‣ Team members who stick with the task at hand may experience stress, particularly as they don't have the support of established processes, or strong relationships with their colleagues.

 CAPE PROJECT MANAGEMENT, INC.

Notes:

Norming

▸ This is when people start to resolve their differences, appreciate colleagues' strengths, and respect the leader.

▸ Now that the team members know one another better, they may socialize together, and they are able to ask each other for help and provide constructive feedback.

▸ People develop a stronger commitment to the team goal, and start to see good progress towards it.

▸ There is often a prolonged overlap between storming and norming, because, as new tasks come up, the team may lapse back into behavior from the storming stage.

CAPE PROJECT MANAGEMENT, INC.

Notes:

Performing

- ▸ The team reaches the performing stage when hard work and minimal friction leads to the achievement of the team's goal.
- ▸ The structures and processes set up support this well.
- ▸ The leader can concentrate on developing individual team members.
- ▸ It feels easy to be part of the team at this stage, and people who join or leave won't disrupt performance.

CAPE PROJECT
MANAGEMENT, INC.

Notes:

DISCUSSION

Have you ever been on a high-performing team? What were some of the attributes?

CAPE PROJECT
MANAGEMENT, INC.

Notes:

Attributes of a High-performing Team

- People have solid and deep trust in each other and in the team's purpose -- they feel free to express feelings and ideas.
- Everybody is working toward the same goals.
- Team members are clear on how to work together and how to accomplish tasks.
- Everyone understands both team and individual performance goals and knows what is expected.
- Team members actively diffuse tension and friction in a relaxed and informal atmosphere.
- The team engages in extensive discussion, and everyone gets a chance to contribute -- even the introverts.

Excerpted from The Collaboration Imperative by Ron Ricci and Carl Wiese.

61

Copyrighted material. 2018

CAPE PROJECT MANAGEMENT, INC.

Notes:

Team Communication in Agile

- Face-to-face
- Osmotic communication
- Open Spaces
- Distributed teams

CAPE PROJECT
MANAGEMENT, INC.

Notes:

Face-to-Face Communication

▸ "The most efficient and effective method of conveying information to and within a development team is face-to-face conversation." – the Agile Manifesto

▸ Co-located teams benefits from this type of communication

▸ User story creation with team members present is highly desirable

▸ Providing feedback during the Sprint would be desirable face-to face

CAPE PROJECT MANAGEMENT, INC.

Notes:

Face-To-Face Communication

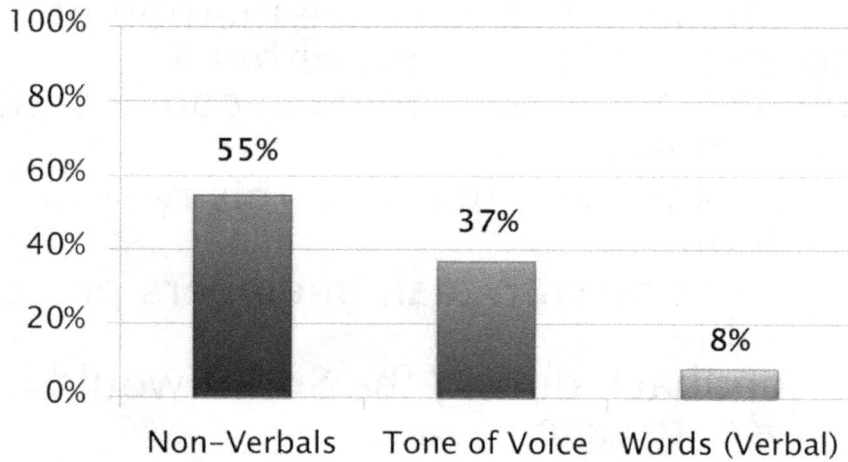

Bar chart showing: Non-Verbals 55%, Tone of Voice 37%, Words (Verbal) 8%

CAPE PROJECT
MANAGEMENT, INC.

Notes:

Osmotic Communication

- ▸ The basis of the open workspace model
- ▸ Information flows into the background hearing of team members
- ▸ Pick up relevant information as though by osmosis
- ▸ Supports fewer meetings
- ▸ Studies show double the productivity and decreased time to market

65

CAPE PROJECT
MANAGEMENT, INC.

Notes:

| |
| |
| |
| |
| |
| |
| |
| |
| |

Open Space Design

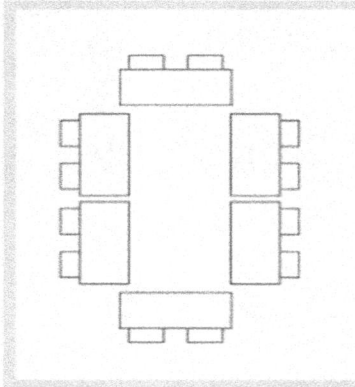

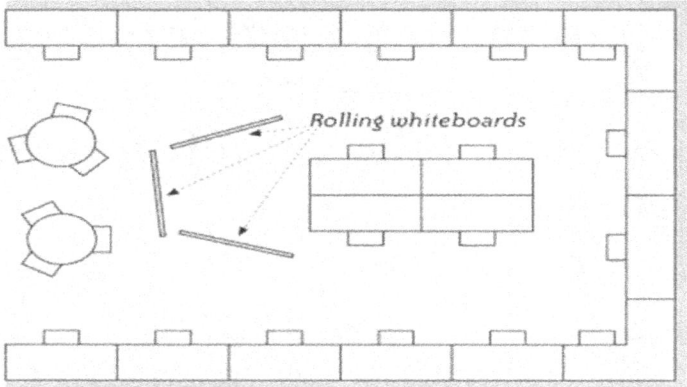

Rolling whiteboards

CAPE PROJECT
MANAGEMENT, INC.

Notes:

Distributed Teams

- ‣ Leverage technology
  - ◦ Email
  - ◦ SMS/instant messaging
  - ◦ Video conferencing
  - ◦ Interactive white boards
  - ◦ Collaboration tools
- ‣ Commit to more informal communications

Copyrighted material. 2018

CAPE PROJECT
MANAGEMENT, INC.

Notes:

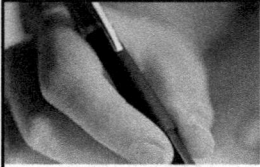

Activity

Scrum Ownership

CAPE PROJECT
MANAGEMENT, INC.

Notes:

Exercise: Scrum Ownership

	Team	Product Owner	Scrum Master
Provides Estimates			
Prioritizes Backlog			
Creates User Stories			
Commits To The Sprint			
Performs User Acceptance			
Facilitates Meetings			
Champion Of Scrum			
Volunteers For Tasks			
Makes Technical Decisions			
Designs Software			
Removes Impediments			
Assign Tasks			

The Scrum Framework

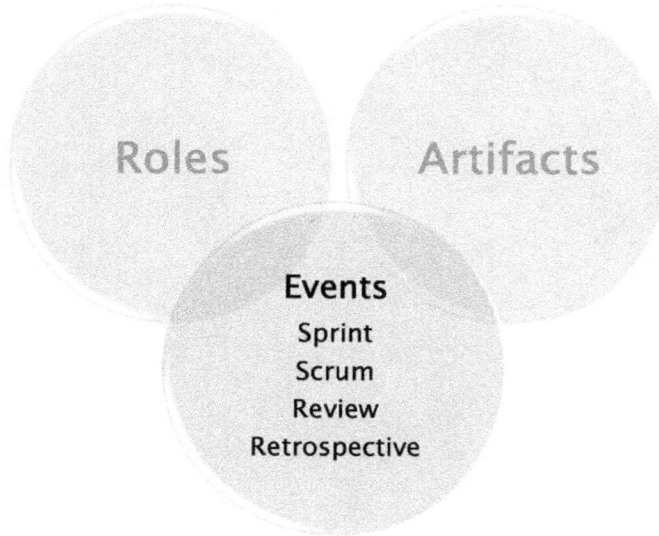

Roles

Artifacts

Events
Sprint
Scrum
Review
Retrospective

CAPE PROJECT
MANAGEMENT, INC.

Notes:

Scrum Events

Agile Discovery occurs before the project begins. It is owned by the Product owner. The goal is to create a high level product roadmap and enough requirements to give the Team work for two to three Sprints.

The **Daily Scrum** is 15 Minutes every day. Same time, same place.

Sprints are a maximum length of one-month. All work needs to meet a definition of "Done".

Scrum

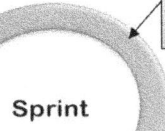

During each **Sprint Review,** "Done" work is shown to stakeholders.

Product Backlog

Sprint Backlog

Sprint

Working Software

Backlog Grooming can take up to 10% of the each Sprint. This involves creating new requirements and prioritizing and estimating them.

Sprint Planning occurs every Sprint. Half the session is to review requirements and half is for design.

Sprint Retrospectives ensure the Team is focusing on continuous improvement.

71

MANAGEMENT, INC.

Notes:

The Scrum Events

Events	Timebox
Sprint Planning	8 hours for a one-month sprint • 1 hour/week of Sprint on requirements • 1 hour/week of Sprint on design
Sprints	One-month maximum
Daily Scrum	15 minutes daily
Backlog Refinement (grooming)	2 hours per week
Sprint Review	4 hours for a one-month sprint • 1 hour/week of Sprint • 1 hour prep
Sprint Retrospectives	3 hours for a one-month sprint
Scrum of Scrums	2-3 times per week for 30 minutes

CAPE PROJECT MANAGEMENT, INC.

Notes:

Why Use Timeboxing?

- ‣ Eliminates the risk of schedule slippage
  - ◦ Time is the fixed constraint
- ‣ Creates a sense of urgency
- ‣ Forces Team to focus on first things first
- ‣ Increases motivation
- ‣ Stops procrastination
- ‣ Creates a working rhythm

73

CAPE PROJECT
MANAGEMENT, INC.

Notes:

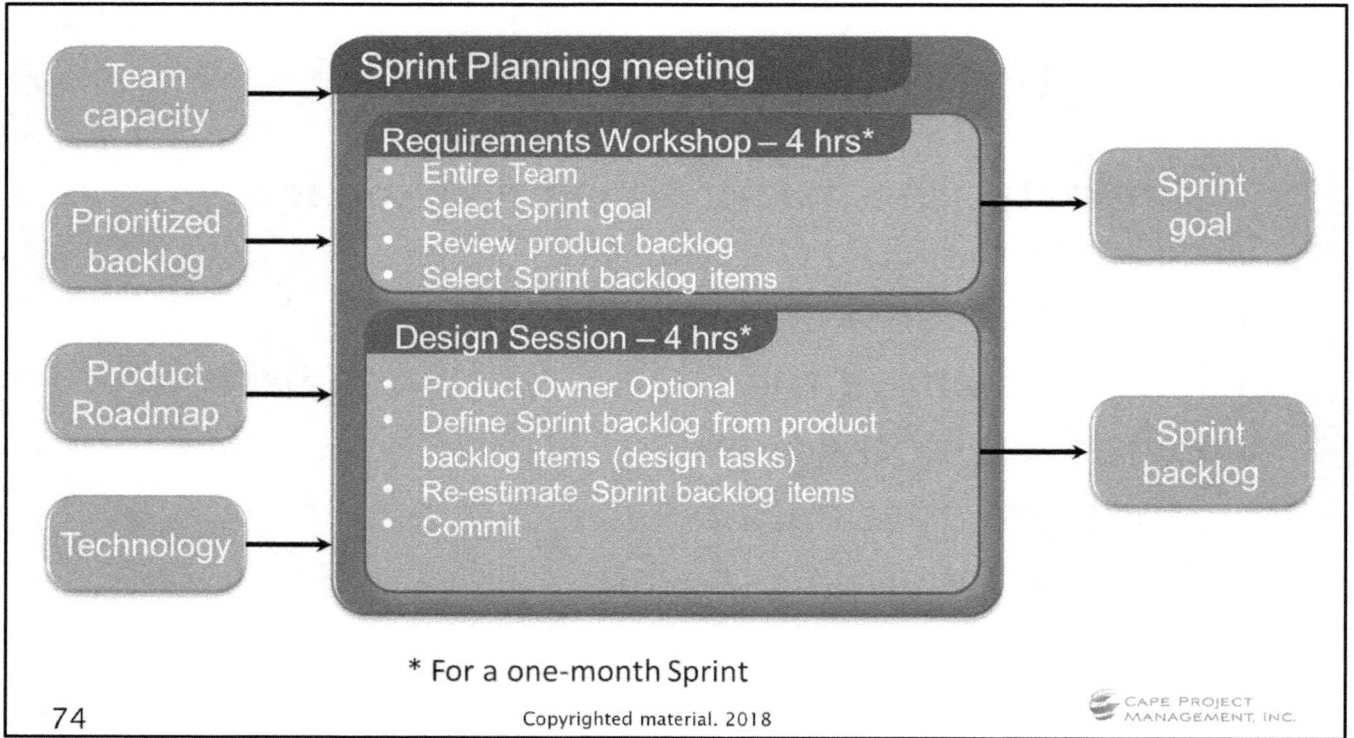

Team capacity →

Prioritized backlog →

Product Roadmap →

Technology →

Sprint Planning meeting

Requirements Workshop – 4 hrs*
- Entire Team
- Select Sprint goal
- Review product backlog
- Select Sprint backlog items

Design Session – 4 hrs*
- Product Owner Optional
- Define Sprint backlog from product backlog items (design tasks)
- Re-estimate Sprint backlog items
- Commit

→ **Sprint goal**

→ **Sprint backlog**

* For a one-month Sprint

74

CAPE PROJECT
MANAGEMENT, INC.

Notes:

Effective Sprint Planning Meetings

- ▸ Start the meeting with a roadmap review
- ▸ The Product Owner needs to be able to articulate a Sprint objective
- ▸ The team decides when the User Stories (requirements) are "good enough" e.g. Definition of Ready
- ▸ Backlog refinement is performed in advance of at least two Sprints

75

CAPE PROJECT
MANAGEMENT, INC.

Notes:

Backlog Refinement

- Each Sprint, the team and Product Owner collaborate to add detail, estimates, and order to items in the Product Backlog.
- Refinement usually consumes no more than 10% of the capacity of the Development Team. Unique to Agile
- Product Owner, Scrum Master, and Dev Team decides how and when refinement is done.
- The Product Backlog items can be updated at any time by the Product Owner

CAPE PROJECT
MANAGEMENT, INC.

Notes:

The Sprint

- ‣ Scrum projects progress via a series of Sprints. Sprints are timeboxed to no more than one month.
- ‣ During a Sprint, the Scrum team takes a small set of features from idea to coded and tested functionality.
- ‣ At the end of the Sprint, these features are done; coded, tested and integrated into the evolving product or system.

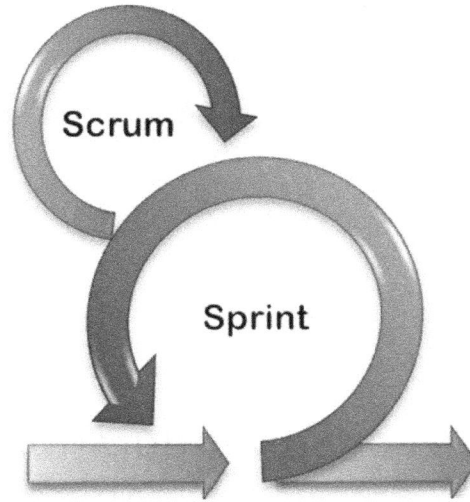

77

CAPE PROJECT
MANAGEMENT, INC.

Notes:

| |
| |
| |
| |
| |
| |
| |
| |

No changes during a Sprint

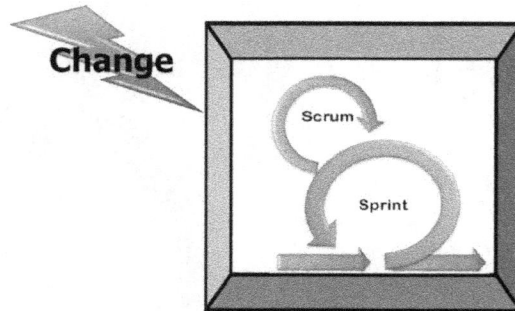

Change

Scrum

Sprint

‣ Plan Sprint durations around how long you can commit to keeping change out of the Sprint

CAPE PROJECT
MANAGEMENT, INC.

Notes:

Four-week Sprint

Pros:
- **Very easy to roadmap.** Long-term, one-year roadmaps are much easier when you only have to plan 12 iterations.
- **Low process load.** The key Scrum meetings only once per month, your team is spending less time in meetings and more time building.

Cons:
- **Long turnaround time.** A new idea that comes in the day after a Sprint kickoff can't be started for a month and wouldn't be demoed until a month after that. Two months turnaround time from request to demo can feel interminable.
- **Mini-waterfall.** There is a risk on a four-week Sprint to break up the weeks into design, develop, test, integrate.

Copyrighted material. 2018

CAPE PROJECT
MANAGEMENT, INC.

Notes:

One-week Sprint

Pros:

▸ **Fast turnaround time.** At most, a new idea has to wait just a week to start development and a week after that for the first product output.

▸ **High energy.** One-week Sprints are fun. The energy is high because the deadline is always this Friday. Every week is an end-of-Sprint rush to the finish line.

Cons:

▸ **Minimal feedback.** There isn't time for a lot of customer feedback before delivering code.

▸ **Lack of a roadmap.** When your horizon is a week it is difficult planning out a year.

▸ **Story sizing.** It can be difficult to size all Stories to fit into a week.

▸ **Relatively heavy...or no process.** A one-week Sprint still needs estimation, tasking, demo, refinement, etc. Relative to the size of the Sprint those fixed time costs will now be high. In practice, the Sprint processes get rushed or omitted.

CAPE PROJECT
MANAGEMENT, INC.

Notes:

DISCUSSION

What is the right Sprint length for you organization?

CAPE PROJECT
MANAGEMENT, INC.

Notes:

The Daily Scrum

- Parameters
  - Daily
  - Same time
  - 15-minutes
  - Stand-up
- Not for problem solving
  - Whole world is invited
  - Only team members and Scrum Master can talk
- Helps avoid other unnecessary meetings

Copyrighted material. 2018 CAPE PROJECT MANAGEMENT, INC.

Notes:

Purpose of the Daily Scrum

- ▸ To help start the day well
- ▸ To support improvement
- ▸ To reinforce focus on the right things
- ▸ To reinforce the sense of team
- ▸ To communicate what is going on

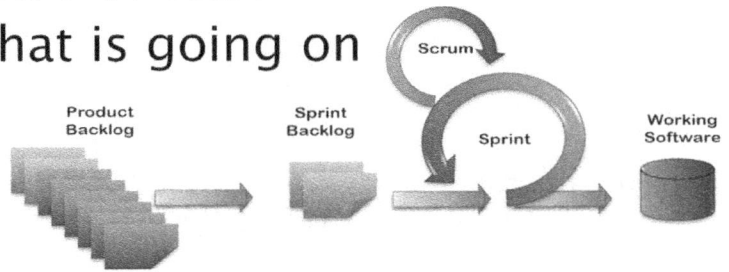

CAPE PROJECT
MANAGEMENT, INC.

Notes:

Everyone answers 3 questions

What did you do yesterday? 1

What will you do today? 2

Is anything in your way? 3

▸ These are *not* a status for the Scrum Master
  ◦ They are commitments in front of peers

CAPE PROJECT MANAGEMENT, INC.

Notes:

Sprint Review Meeting

‣ Time-boxed—one hour per each week of Sprint
‣ One-hour prep
‣ Only "done" items are demonstrated
  ◦ Done versus done-done
‣ Opportunity for "inspect and adapt"
‣ Primary audience are stakeholders
‣ Feedback is encouraged

CAPE PROJECT
MANAGEMENT, INC.

Notes:

Sprint Review Do's and Don'ts

Do's
- Ensure it is the Product Owner's meeting
- Show the progress against the Product Backlog & Roadmap
- Make certain stakeholders are present
- Prepare in advance
- Team members present their own work

Don'ts
- The Product Owner acts as customer (they should not be seeing anything for the first time)
- Allow Sprint Reviews to become boring meetings
- Not showing working and tested software (no PPTs)

CAPE PROJECT MANAGEMENT, INC.

Notes:

Sprint Retrospective Meeting

- ▸ Periodically take a look at what is and is not working
- ▸ Only attended by the Scrum Team (no stakeholders)
- ▸ Done after every Sprint
- ▸ Time-boxed: 3 hours for a one month Sprint
- ▸ Use a neutral facilitator
- ▸ Important to have open feedback
- ▸ Incorporate feedback into next Sprint

87

CAPE PROJECT
MANAGEMENT, INC.

Notes:

Sprint Retrospectives

- ‣ Examines the way work was performed
- ‣ Inspects how the last Sprint went in terms of people, relationships, process, and tools
- ‣ Identifies and prioritizes the major items that went well as well as potential improvements
- ‣ Creates a plan for implementing improvements to the way the Team does its work

88

CAPE PROJECT
MANAGEMENT, INC.

Notes:

Start / Stop / Continue

▸ Whole team gathers and discusses what they'd like to:

Start doing

Stop doing

This is just one of many ways to do a Sprint Retrospective.

Continue doing

Copyrighted material. 2018

CAPE PROJECT
MANAGEMENT, INC.

Notes:

DISCUSSION

Which ritual is easiest?
Which is hardest?

CAPE PROJECT
MANAGEMENT, INC.

Notes:

Sample end of Sprint agenda – The "Planning Day"

Time	Item (two week Sprint)
9:00–9:15	Daily Scrum • Last Scrum of prior Sprint–discuss status of any incomplete stories
9:15–10:30	Sprint Review: • Demo of stories delivered over the course of last Sprint
10:30–11:30	Sprint Retrospective • Focus both on product and process opportunities
11:30–12:30	Lunch
12:30 – 2:30	Sprint Planning Requirements Session: • Review roadmap, discuss sprint goal, review each story in priority order and estimate until velocity is met
2:30–4:30	Sprint Planning Design Session: • Team reviews stories, creates tasks for each
4:30–4:45	Team commits to the Sprint Backlog • Ready to start with a daily Scrum the next day

CAPE PROJECT
MANAGEMENT, INC.

Notes:

The Scrum Framework

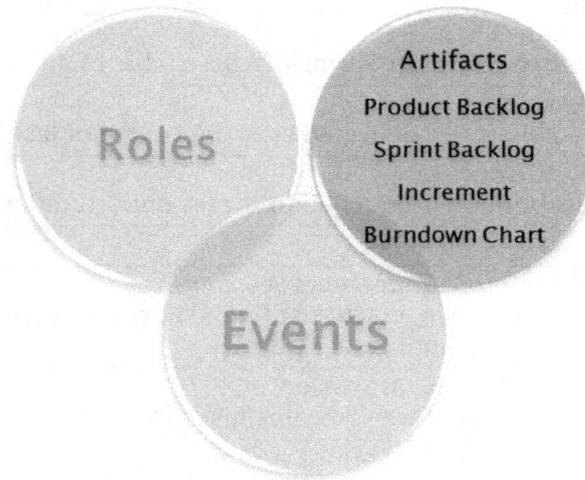

Roles

Events

Artifacts
Product Backlog
Sprint Backlog
Increment
Burndown Chart

92

CAPE PROJECT
MANAGEMENT, INC.

Notes:

Product Backlog

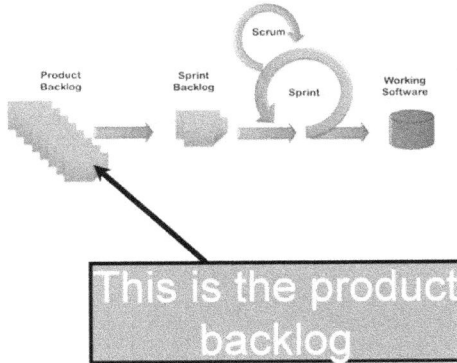

This is the product backlog

- ‣ The requirements
- ‣ A list of all desired work on the project
- ‣ Ideally expressed such that each item has value to the users or customers of the product
- ‣ Prioritized by the product owner
- ‣ Reprioritized at the start of each Sprint

93

CAPE PROJECT MANAGEMENT, INC.

Notes:

Product Backlog

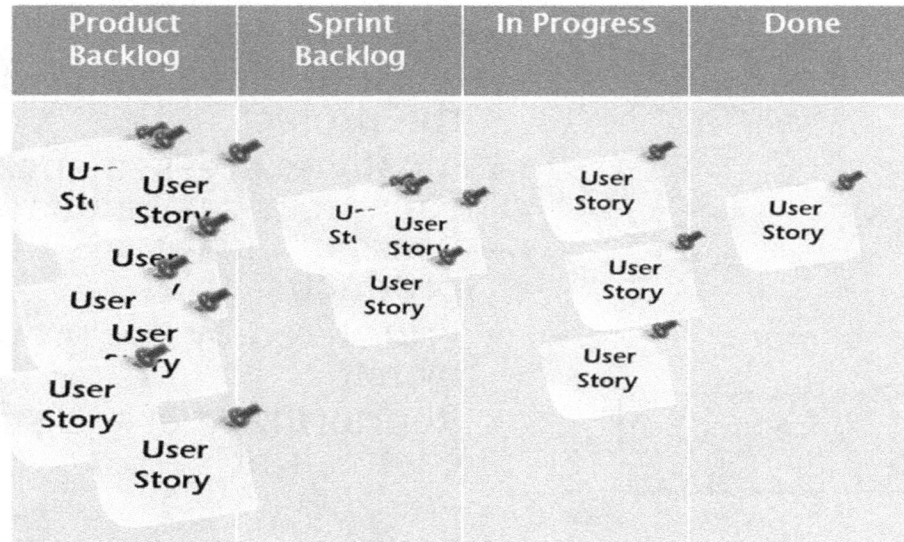

Product Backlog	Sprint Backlog	In Progress	Done
User Story	User Story	User Story	User Story
User Story	User Story	User Story	
User Story		User Story	
User Story			
User Story			
User Story			

CAPE PROJECT MANAGEMENT, INC.

Notes:

The Sprint backlog

- ‣ The Product Owner works with the development team to select Stories based upon a Sprint Goal
- ‣ Individuals sign up for work of their own choosing
- ‣ Any team member can add, delete or change tasks on the Sprint backlog
- ‣ Estimated work remaining is updated daily on backlog and burndown charts by the development team.

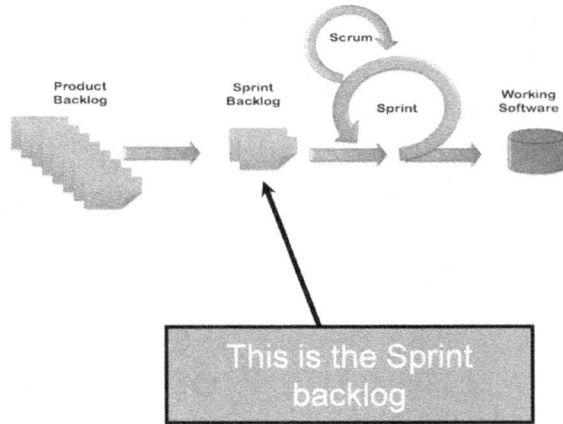

Product Backlog

Sprint Backlog

Scrum

Sprint

Working Software

This is the Sprint backlog

CAPE PROJECT MANAGEMENT, INC.

Notes:

| |
| |
| |
| |
| |
| |
| |
| |

An Increment in Scrum*

▸ The Increment is the sum of all the Product Backlog items completed during a Sprint and the value of the increments of all previous Sprints.

▸ At the end of a Sprint, the new Increment must be "Done," which means it must be in useable condition and meet the Scrum Team's definition of "Done."

▸ It must be in useable condition regardless of whether the Product Owner decides to actually release it.

*The "third" artifact in the Scrum Guide

96   Copyrighted material. 2018   CAPE PROJECT
                                                                          MANAGEMENT, INC.

Notes:

Definition of "Done"

- The list of activities (coding comments, unit testing, integration testing, release notes, design documents, etc.) which supports the expected business value.
- The intention is to focus on value-added steps that allow the team to focus on what must be completed in order to build software while eliminating wasteful activities.
- The definition of "Done" is defined by the Product Owner and Development Team.
- Make "Done" more stringent over time with each release

97

CAPE PROJECT
MANAGEMENT, INC.

Notes:

Scrum Training for Teams

Burndown Charts

▸ Tracks work remaining
▸ At–a–glance information
   ◦ Number of user stories committed
   ◦ Duration of Sprint
   ◦ Target velocity
   ◦ Performance against plan
      · Above the line– behind schedule
      · Below the line – ahead of schedule

98

Copyrighted material. 2018

CAPE PROJECT MANAGEMENT, INC.

Notes:

Velocity

- ▸ Velocity = Rate of Progress
- ▸ Number of story points to be completed in Sprint
- ▸ First Sprint is a guess
- ▸ Estimate improves over time
- ▸ Account for work done and disruptions on the project
- ▸ Based upon team synergy

99

CAPE PROJECT
MANAGEMENT, INC.

Notes:

Sprint Burndown Chart
Work Remaining

Beginning of Iteration

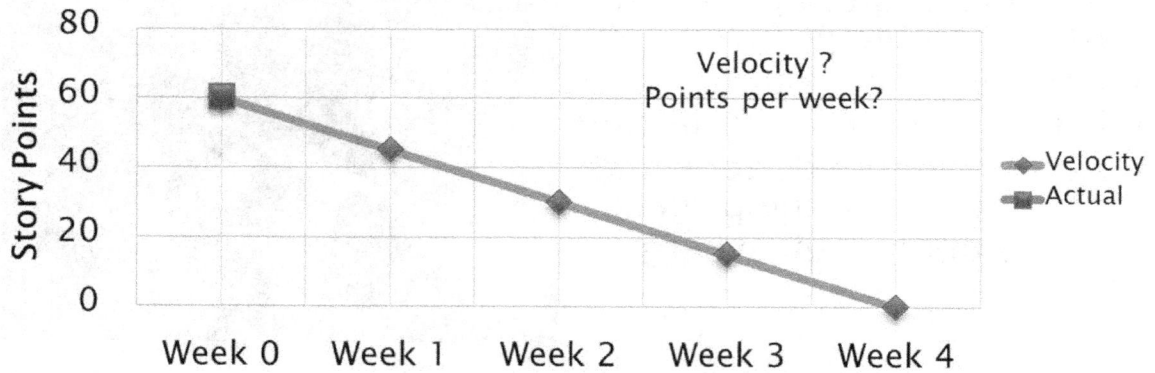

Velocity ?
Points per week?

Story Points: 0, 20, 40, 60, 80

Week 0, Week 1, Week 2, Week 3, Week 4

Velocity
Actual

CAPE PROJECT MANAGEMENT, INC.

Notes:

Sprint Burndown Chart

End of Week 1

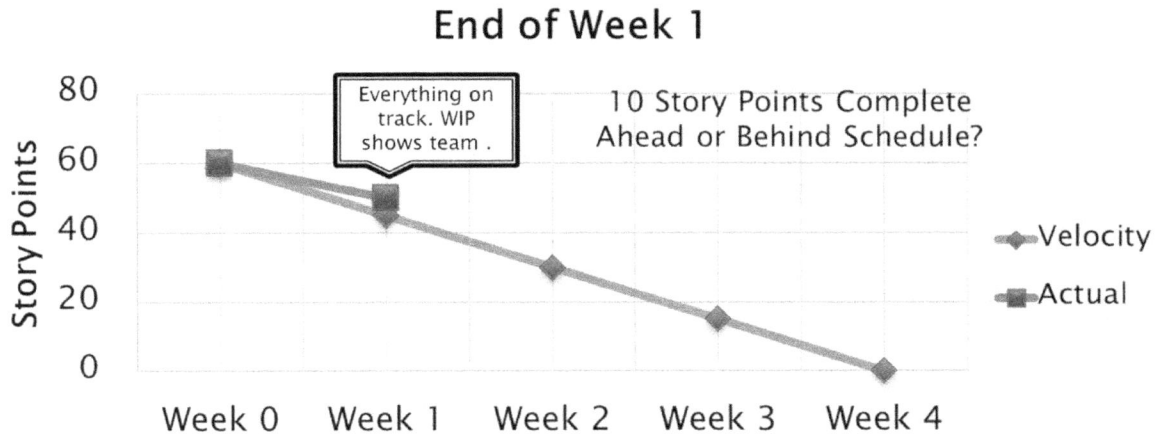

Everything on track. WIP shows team .

10 Story Points Complete
Ahead or Behind Schedule?

CAPE PROJECT
MANAGEMENT, INC.

Notes:

Sprint Burndown Chart

End of Week 2

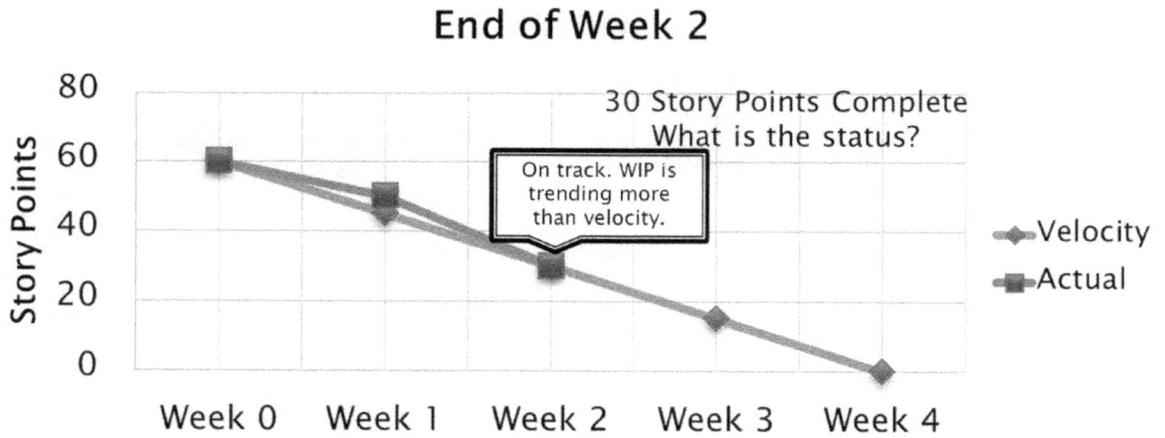

30 Story Points Complete
What is the status?

On track. WIP is trending more than velocity.

Velocity
Actual

Copyrighted materials. 2018

CAPE PROJECT
MANAGEMENT, INC.

Notes:

Sprint Burndown Chart

End of Week 3

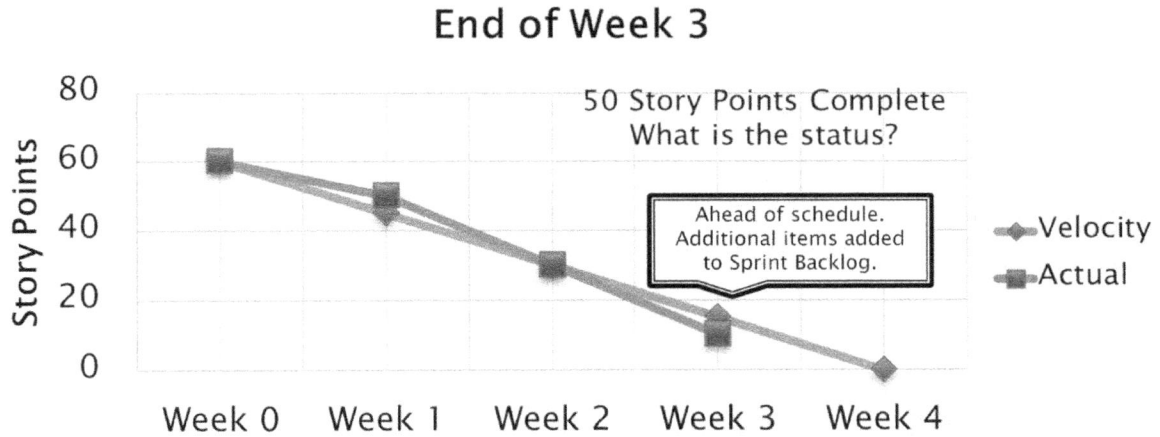

50 Story Points Complete
What is the status?

Ahead of schedule.
Additional items added
to Sprint Backlog.

Velocity
Actual

CAPE PROJECT
MANAGEMENT, INC.

Notes:

Sprint Burndown Chart

End of Iteration/Sprint

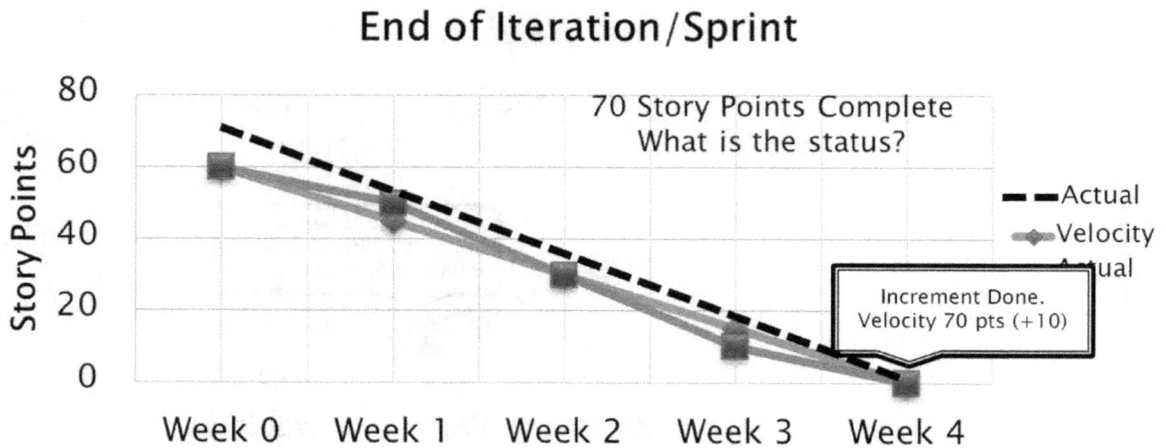

70 Story Points Complete
What is the status?

- - - Actual
Velocity
Actual

Increment Done.
Velocity 70 pts (+10)

CAPE PROJECT
MANAGEMENT, INC.

Notes:

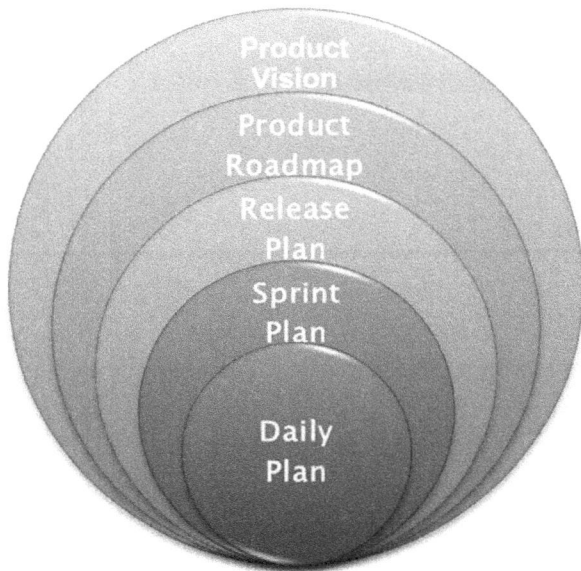

The Agile Product Lifecycle

Product Vision
Product Roadmap
Release Plan
Sprint Plan
Daily Plan

Copyrighted material. 2018

CAPE PROJECT
MANAGEMENT, INC.

CAPE PROJECT MANAGEMENT, INC.

Notes:

Agile Planning Approach

Product or Project

What business objectives will the product fulfill?

Product Vision

Product Roadmap

Iteration/Sprint

What specifically will we build? (User Stories)

How will this iteration move us toward release objectives?

Sprint Goal

Sprint Backlog

Release

How can we release value incrementally?

What subset of business objectives will each release achieve?

What users will the release serve?

What general capabilities will the release offer?

Release plan

Product Backlog

Story (Backlog Item)

What user or stakeholder need will the story serve?

How will it specifically look and behave?

How will I determine if it's completed?

Story Details/Tasks

Acceptance Tests

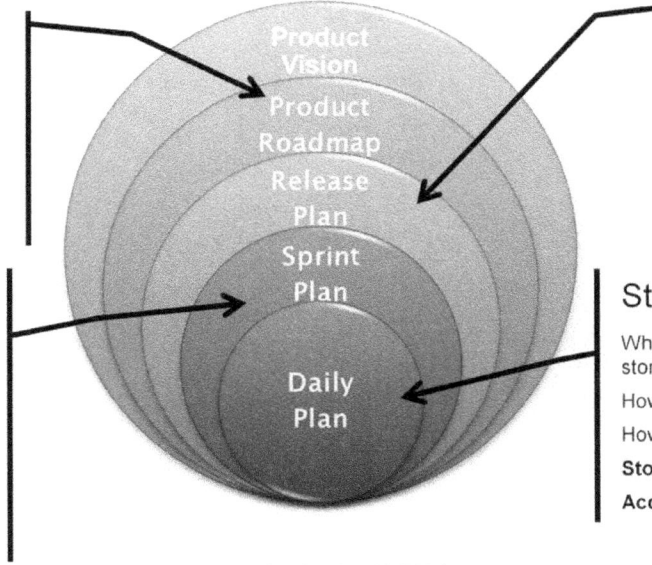

Product Vision
Product Roadmap
Release Plan
Sprint Plan
Daily Plan

107

CAPE PROJECT MANAGEMENT, INC.

Notes:

| |
| |
| |
| |
| |
| |
| |
| |

Agile Chartering

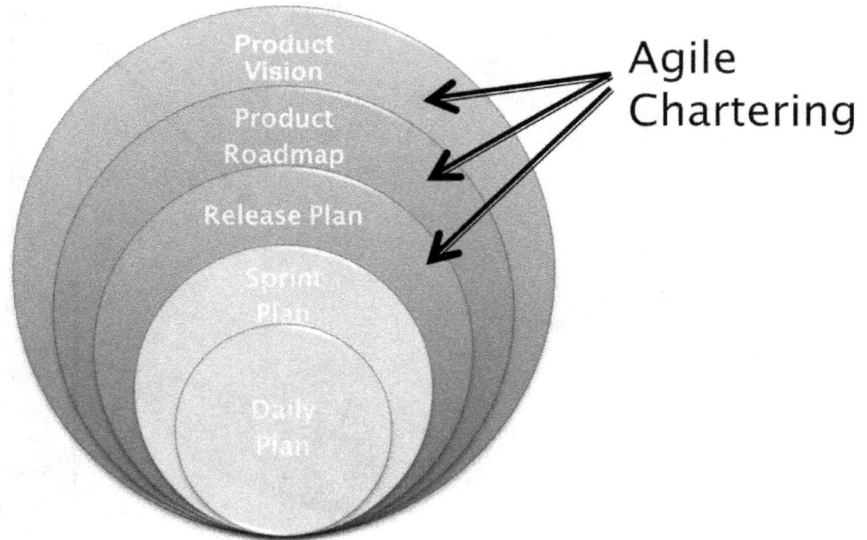

Product Vision
Product Roadmap
Release Plan
Sprint Plan
Daily Plan

Agile Chartering

 CAPE PROJECT MANAGEMENT, INC.

Notes:

Developing a Vision

- ‣ The product vision is key to the success of the project.
- ‣ The product vision should align with the company vision
- ‣ The vision should be revisited frequently
- ‣ All releases of the product should related back to the vision

CAPE PROJECT
MANAGEMENT, INC.

Notes:

Create a Vision Board

Vision: What is your overarching goal for creating the product?			
Product	Needs	Target Group	Value
What is the product? What makes it desirable? Is it feasible to develop?	What problem does the product solve? What benefit does it provide?	Who are the target users and customers?	How is the product going to benefit the company? What are the business goals?

110

Notes:

Vision Board Example

Vision: The Learning Management System (LMS) will enhance the educational experience by giving students and educators more ways to stay engaged online – both in and outside of the classroom. It will give students and faculty access to their courses, content, and grading and will allow them to participate in an online learning community on their desktop and variety of mobile devices.

Product	Needs	Target Group	Value
Online access to courses, grading, and student profiles. Eliminates paper and provides platform for collaboration. Development team and expertise exists.	Eliminates the need for paper catalogs, add/drop forms and high involvement of faculty. Reduces overhead, 24/7 access and ability to change content.	Students Faculty Administrators Parents	Customer base is every university, school, and training organization in the world. Generate revenue with each release of functionality.

Notes:

Product Roadmap

- A roadmap is a planned future, laid out in broad strokes
  - Planned or proposed product releases, listing high level functionality or release themes, laid out in rough timeframes
  - For a period usually extending for 2 or 3 significant feature releases into the future
- Shows progress towards strategy
- Lots of "wiggle room"
- Example:
  - Implement course listing functionality
  - Implement grading functionality
  - Implement discussion groups
  - Implement student profiles

112

CAPE PROJECT MANAGEMENT, INC.

Notes:

Product Roadmap Example

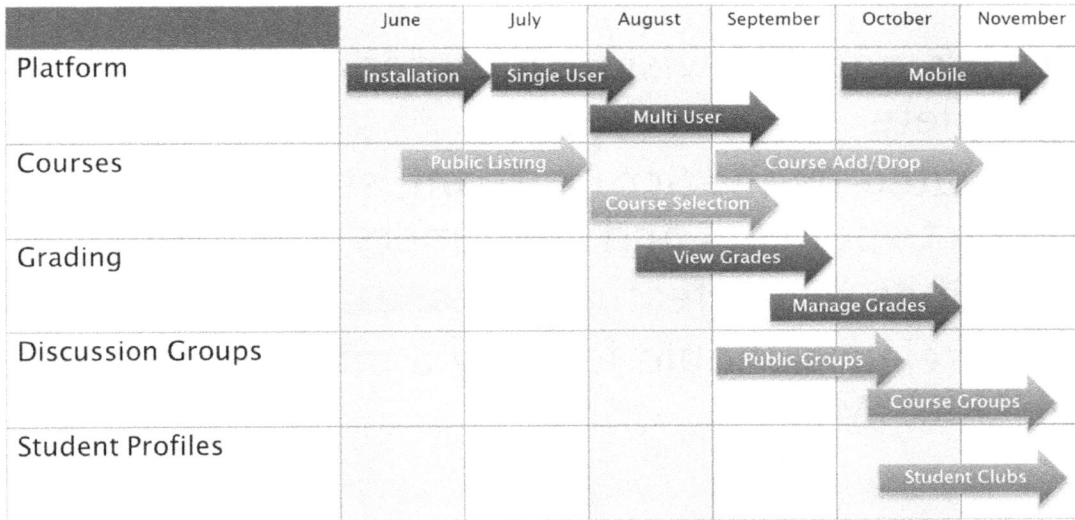

	June	July	August	September	October	November
Platform	Installation	Single User	Multi User		Mobile	
Courses	Public Listing		Course Selection	Course Add/Drop		
Grading			View Grades	Manage Grades		
Discussion Groups				Public Groups	Course Groups	
Student Profiles					Student Clubs	

113

Copyrighted material. 2018

CAPE PROJECT MANAGEMENT, INC.

Notes:

Release Planning

- ‣ Occurs once the vision and roadmap are complete
- ‣ Is created by the Scrum Team, stakeholders, project sponsors and customers when possible
- ‣ Identifies major feature releases for 3–6 months
- ‣ Each release should identify a minimum viable product (MVP)

114

CAPE PROJECT
MANAGEMENT, INC.

Notes:

Minimum Viable Product (MVP)

- ▸ The product with the highest return on investment versus risk.
- ▸ Just those core features that allow the product to be deployed, and no more.
- ▸ It allows you to test an idea by exposing an early version of your product to the target users and customers, to collect the relevant data, and to learn from it.

Source: http://www.romanpichler.com/blog/minimum-viable-product-and-minimal-marketable-product/

115

Copyrighted material. 2018

CAPE PROJECT MANAGEMENT, INC.

Notes:

115

A Product Release Plan

- ‣ Goes into next level of detail
- ‣ Sets a common understanding
- ‣ A projection, not a commitment
- ‣ Example
  - ◦ Release 1:
    - · LMS Installed with pilot group logins validated
    - · Pilot with 3 faculty and 60 students
    - · Course listings available
  - ◦ Release 2
    - · Incorporate pilot feedback
    - · Enable College of Engineering faculty and students
    - · Implement course selection on-line
  - ◦ Release 3
    - · ...

CAPE PROJECT
MANAGEMENT, INC.

Notes:

Requirements Management

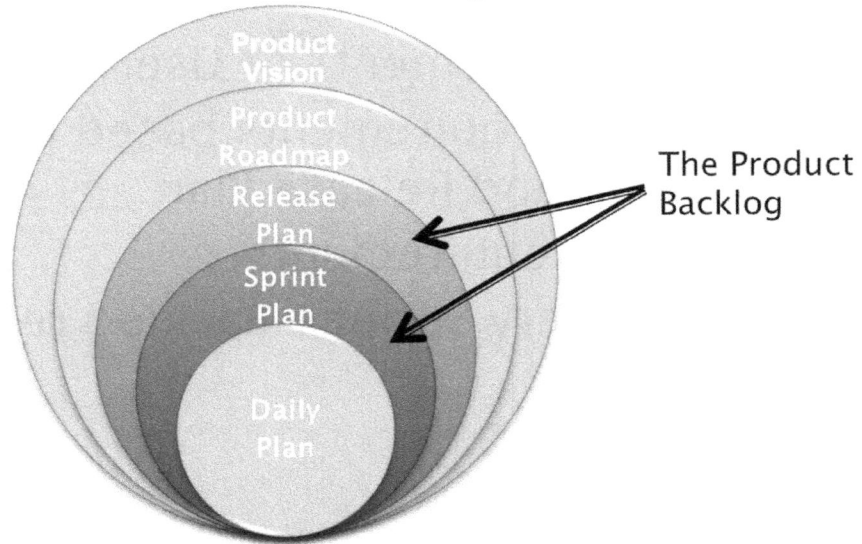

The Product Backlog

CAPE PROJECT
MANAGEMENT, INC.

Notes:

Creating a Product Backlog

1. Create requirements or User Stories
2. Prioritize Requirements based upon Roadmap and Release Plan
3. Estimate Requirements
4. Update every Sprint – Backlog refinement

118

CAPE PROJECT
MANAGEMENT, INC.

Notes:

Types of Backlogs

Type	Definition
Product	Set of prioritized requirements aligning with product vision
Release	The minimum requirements that would support a release. • Smallest possible release contains one Sprint • "Done–Done"
Sprint	A subset of requirements selected according to the velocity (capacity) of the team

CAPE PROJECT
MANAGEMENT, INC.

Notes:

Key Principles for Agile Requirements

- Active user involvement is imperative
- Agile teams must be empowered to make decisions
- Requirements emerge and evolve as software is developed
- Agile requirements are 'barely sufficient'
- Requirements are developed in small pieces
- Enough is enough – apply the 80/20 rule
- Cooperation, collaboration and communication between all team members is essential
- All requirements as captured in as product backlog items (PBIs)
- User Stories are one technique to capturing functional requirements in Agile

120

CAPE PROJECT
MANAGEMENT, INC.

Notes:

Notes:

Agile Requirements – User Stories

- Agreement between customer and developer to have a conversation.
- A User Story is a very high-level definition of a requirement.
- Contains just enough information to produce a reasonable estimate of the effort to implement it.
- Product Owner writes User Stories on behalf of customer
  - Written in language of business to allow prioritization
  - Customer is primary product visionary

CAPE PROJECT MANAGEMENT, INC.

Notes:

User Stories have 3 parts

Card *What is the goal of a user*	**As a (user role), I want to (goal) so I can (reason)** *Example:* *As a registered student, I want to view course details so I can create my schedule*
Conversation *How to achieve the goal using the system?*	**Discuss the card with a stakeholder. Just in time analysis (JIT) through conversations.** *Example:* *What information is needed to search for a course?* *What information is displayed?*
Confirmation *How to verify if the story is done and complete, and the goal is achieved*	**Record what you learn in an acceptance test.** *Example:* *Student can access course catalog 24 x 7 hours* *Student cannot choose more than three courses*

http://ronjeffries.com/xprog/articles/expcardconversationconfirmation/

CAPE PROJECT
MANAGEMENT, INC.

Notes:

Why use User Stories?

▸ Keep the focus on expressing business value
▸ Avoids introducing detail too early that would prevent design options and inappropriately lock developers into one solution
▸ Avoids the appearance of false completeness and clarity
▸ Invites negotiation and movement in the backlog
▸ Leaves the technical functions to the architect, developers, testers, and so on

124   Copyrighted material. 2018   CAPE PROJECT MANAGEMENT, INC.

Notes:

Guidelines for Good Stories

- ‣ Start with themes or epics
- ‣ Create user stories that meet the criteria of INVEST
- ‣ Include user roles in stories rather than saying "user"
- ‣ Don't rely solely on stories if they can be better expressed in other ways
- ‣ Size your story appropriately for the time frame it may be implemented in

CAPE PROJECT
MANAGEMENT, INC.

Notes:

Themes , Epics and User Stories

Themes

▸ Themes are groups of related stories. Often the stories all contribute to a common goal or are related in some obvious way, such as all focusing on a single function.

Epics

▸ Epics resemble themes in the sense that they are made up of multiple stories. As opposed to themes, however, these stories often comprise a complete workflow for a user.

User Stories

▸ A User story is a self-contained unit of work agreed upon by the developers and the stakeholders. Stories are the building blocks of your sprint.

126

CAPE PROJECT
MANAGEMENT, INC.

Notes:

Agile Requirements Hierarchy

Copyrighted material. 2018

CAPE PROJECT
MANAGEMENT, INC.

Notes:

Themes, Epics and User Stories

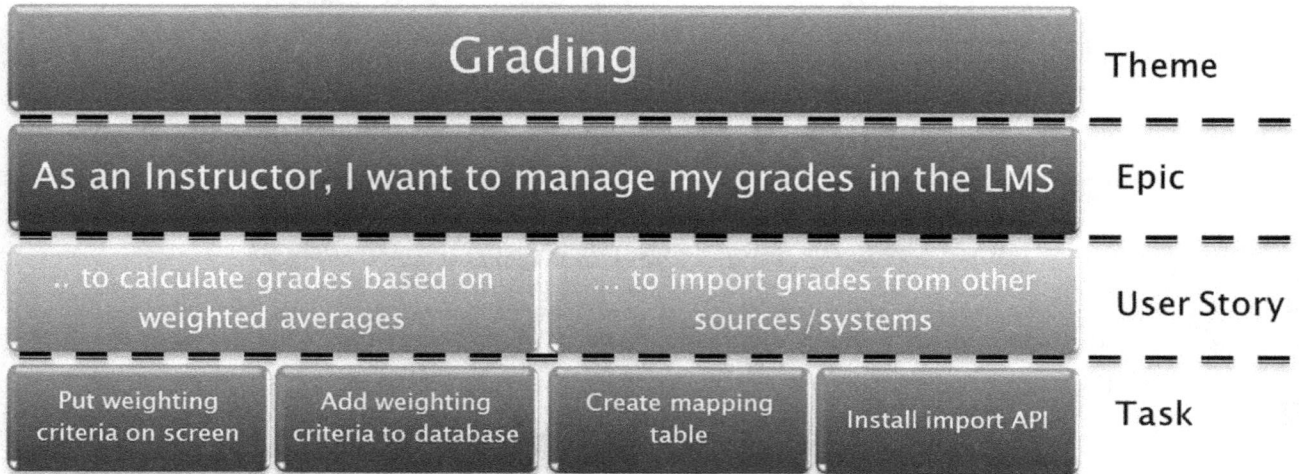

Grading	Theme

As an Instructor, I want to manage my grades in the LMS	Epic

.. to calculate grades based on weighted averages	... to import grades from other sources/systems	User Story

Put weighting criteria on screen	Add weighting criteria to database	Create mapping table	Install import API	Task

CAPE PROJECT
MANAGEMENT, INC.

Notes:

Writing Stories

▸ Good stories are:
- Independent
- Negotiable
- Valuable to users or customers
- Estimable
- Small
- Testable   (INVEST)

129   Copyrighted material. 2018   CAPE PROJECT
                                                                  MANAGEMENT, INC.

Notes:

Independent

- Stories that depend on other stories are difficult to prioritize and estimate
- Dependent Story:
  - As a student, I want to be able to log in and update my profile
- Independent Stories
  - As a student, I want to be able to log in
  - As a student, I want the ability to update my profile

CAPE PROJECT
MANAGEMENT, INC.

Notes:

Negotiable

‣ User Stories serve as reminders not contracts
‣ The story should convey the problem, not the solution. This way the solution is negotiable.
‣ Details need to be fleshed out in conversation

CAPE PROJECT
MANAGEMENT, INC.

Notes:

Valuable

- Both to people using the software and paying for the software
- Avoid stories valued only by developers (make the benefits to customers/users apparent for these stories)
- Example
  - "All connections to the database are through a connection pool" could be rewritten as "Up to 50 users should be able to use the application with a 5-user database license"

CAPE PROJECT
MANAGEMENT, INC.

Notes:

Estimable

- ▸ 3 common reasons why a story might not be estimable
- ▸ Not enough information or team lacks domain knowledge
  - ◦ Get details from customer
- ▸ New technology or not enough knowledge in the team
  - ◦ Perform spike to explore technology
- ▸ Story is too big
  - ◦ Split the story into smaller ones

Copyrighted material. 2018

CAPE PROJECT MANAGEMENT, INC.

Notes:

Small

‣ Easy to use in planning
‣ Split compound & complex stories (Epics)
‣ Combine too small stories

Copyrighted material. 2018 CAPE PROJECT MANAGEMENT, INC.

Notes:

Testable

- Can't tell if story is "done" without tests
- Aim for most tests to be automated
- Include Acceptance Criteria as part of the User Story

CAPE PROJECT
MANAGEMENT, INC.

Notes:

| |
| |
| |
| |
| |
| |
| |
| |
| |

When is a User Story Ready?

- A story is **clear** if all Scrum team members have a shared understanding of what it means.
- An item is **testable** if there is an effective way to determine if the functionality works as expected. Acceptance criteria exists ensure that each story can be tested, typically there are three to five acceptance criteria per User Story.
- A story is **feasible** if it can be completed in one Sprint, according to the definition of done.
- **Ready** stories are the output of the product backlog refinement work.

Note: some organizations create a Definition of "Ready"

CAPE PROJECT
MANAGEMENT, INC.

Notes:

GROUP EXERCISE

Evaluate User Stories

CAPE PROJECT
MANAGEMENT, INC.

Notes:

User Story Examples: LMS Project

Epic 1 Document Management and Editing		Good/Bad
As a Faculty Member, I want...	assignments submitted via the LMS to be searched for direct quotes, so that I can identify students and work with plagiarism.	
As a Faculty Member, I want...	the first release to be in 4 months so that I can use it in the Fall semester	
As a Faculty Member, I want...	to download completed assignments directly to my computer, so that I can edit files outside the LMS.	
As a Faculty Member, I want...	to ensure the Helvetica is part of the font set since it is the most readable font	
As a Student, I want...	to wait to deploy this software until after I graduate so that I don't have to learn something new	
As a Student, I want...	to save content from the LMS as a PDF, so I can protect my content from editing.	
As a Student, I want...	to export the content of a discussion, so that I can view, edit and save in a word processing application on my desktop.	
As the Department Head	I want the deployment of document management to be on time and on budget so that I get my bonus	
Epic 2 Communications		
As a Faculty Member, I want...	to identify specific discussion forum posts, so that I can grade the student responses.	
As a Faculty Member, I want...	to be notified whenever a new message is posted to a discussion I am subscribed to, so that I don't have to enter the LMS to check for each topic's updates.	
As a Student, I want to...	view in one place all classmates that have contributed to a project, document activity, so that I can see what has been done and by whom.	
As a Student, I want to...	view all assignments and due dates for a course in one place, so that I can identify what I need to do and when across the course.	

User Story Examples: LMS Project (cont'd)

As a Student, I want to...	see who else is online across the LMS, so I can communicate with classmates and faculty in real time.	
As a Student, I want to...	to post profile information about me, so other students and faculty can know more about each other.	
Epic 3 Assessment & Grading		
As a Faculty Member, I want...	to import exams/assessments and exam questions from external sources (e.g. as MS-Word documents), so that I don't have to reenter them in the LMS.	
As a Faculty Member, I want...	I want the LMS to be easy to use so that I get good evaluations	
As a Faculty Member, I want...	to calculate grades based on weighted averages, so that students always know their standing in the course.	
As a Faculty Member, I want...	to allow individual students to view individual grades, so that students can see their grades per assignment.	
As a Faculty Member, I want...	to export grades in standard and popular file/exchange formats, so I can use external applications to manage my grades.	
As a Student, I want to...	the option of showing everyone's grades in the class anonymously, so that students can compare their grades with the rest of the class.	
As a Student, I want to...	view all of my grades across all courses, so I so not need to enter each course.	
As a Student, I want to...	I want the courses to be easy so that I don't fail	
As a Trainer, I want...	to use secure exam tools so that I can administer exam to both online and in-person students.	
As a Trainer, I want...	I want the training to be engaging so that the class doesn't fall asleep	

Prioritize items in a backlog

- The Product Owner prioritizes the product backlog items
- The backlog is prioritized so that the most valuable items have the highest priorities
- Prioritization by Release translates to the Product Roadmap

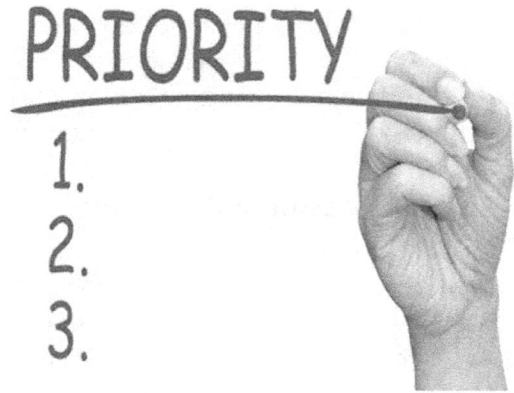

PRIORITY

1.

2.

3.

CAPE PROJECT
MANAGEMENT, INC.

Notes:

Customer-Valued Prioritization Techniques

- ‣ Cumulative voting (the money game)
- ‣ MoSCoW prioritization
- ‣ Kano analysis
- ‣ Risk-based prioritization
- ‣ Weight Shortest Job First (WSJF)
- ‣ Pareto analysis

CAPE PROJECT
MANAGEMENT, INC.

Notes:

Agile Sizing and Estimating

- ▸ Assumes all traditional estimates are inaccurate, essentially we are guessing
- ▸ Focuses on rapid/order of magnitude estimating

CAPE PROJECT
MANAGEMENT, INC.

Notes:

Estimating and Scaling/Sizing Techniques

- ‣ Story points
- ‣ Planning poker
- ‣ T-Shirt Sizes
- ‣ Fibonacci Sequence

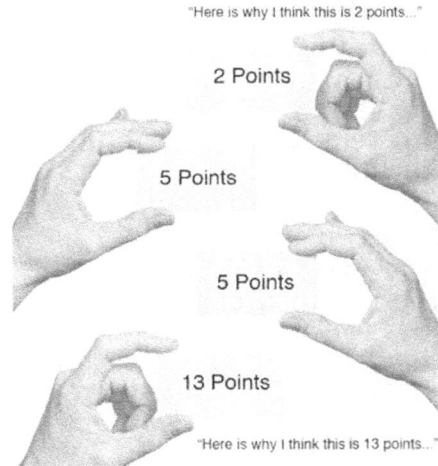

"Here is why I think this is 2 points..."

2 Points

5 Points

5 Points

13 Points

"Here is why I think this is 13 points..."

143

CAPE PROJECT
MANAGEMENT INC.

Notes:

Why use Story Points

- **Quick**: Rather, the goal is to quickly estimate a level of effort, and you can do it much faster than you can when using traditional estimating approaches.
- **Accurate**: By refining the backlog, you resize and develop quick estimates of effort in an Agile manner. The rough classifications of story point relative estimation are a more accurate and flexible way to determine priorities and schedule.
- **Improves over Time**: Over time, you can look at how many points your team typically completes within a sprint, and become better and better at relative estimating.
- **Project-specific**: It's nearly impossible to predict an exact amount of hours for any given story, because hours are relative numbers. Creating a User Story benchmark for a project allows you to attain a more accurate picture of velocity.

CAPE PROJECT MANAGEMENT, INC.

Notes:

Planning Poker

- ▸ Each card represents a size of User Story
- ▸ Everyone should agree on what an average size story is (benchmark)
- ▸ Each story is given a card
- ▸ Play poker on each story until the variance is 1
- ▸ Choose the higher number for the Sprint plan
- ▸ If a User Story is bigger than an iteration or difficult to size use 15, 20, 25, 50 for Jack, Queen, King, Ace (Epic)

145

CAPE PROJECT
MANAGEMENT, INC.

Notes:

T-Shirt Sizes

- Small, Medium, Large, X-Large, XX-Large
  - 1,2,4,8,16 Story Points
- Find your smallest story and give it a 1
- Use cards to estimate
  - Similar to planning poker

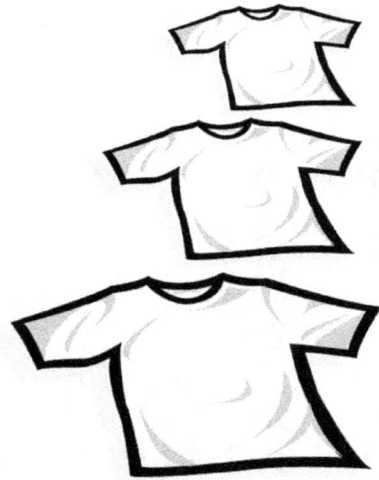

Copyrighted material. 2018

Notes:

Fibonacci Sequence

- 0 1 1 2 3 5 8 13 21 34 55...
- Provides more flexibility for unknown larger stories
- Find your smallest story and give it a 1
- Limit the largest number
- Use index cards and estimate
  - similar to planning poker

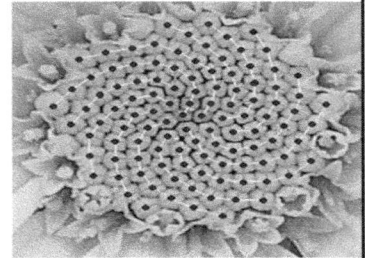

CAPE PROJECT
MANAGEMENT, INC.

Notes:

Relative Estimating using Swimlanes

1	2	3	5	8	13	~

‣ Order each user story from easiest to hardest

‣ Group similar size stories together

‣ Leave a column for Epics

‣ Label using Fibonacci last

http://theagilepirate.net/archives/109

CAPE PROJECT
MANAGEMENT, INC.

Notes:

Assigning Story Points

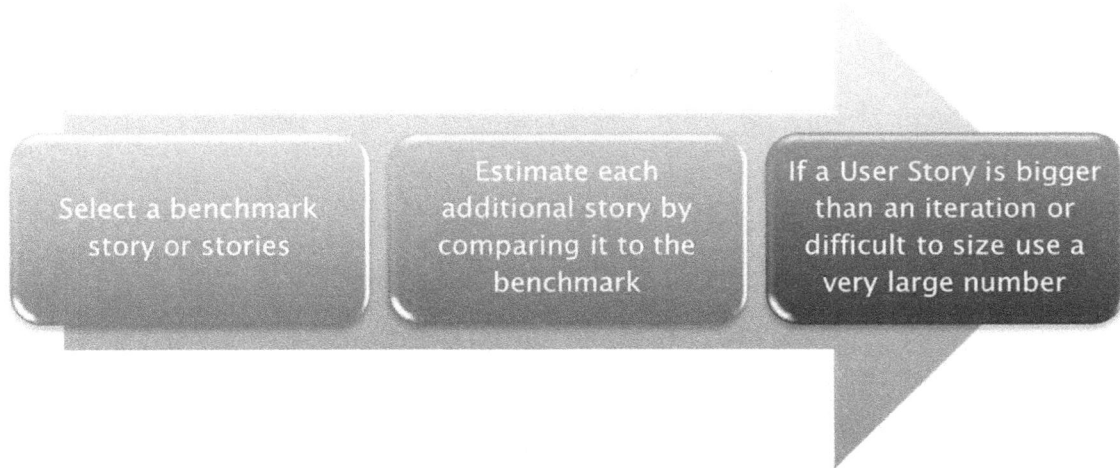

Select a benchmark story or stories

Estimate each additional story by comparing it to the benchmark

If a User Story is bigger than an iteration or difficult to size use a very large number

CAPE PROJECT MANAGEMENT, INC.

Notes:

Sprint Planning

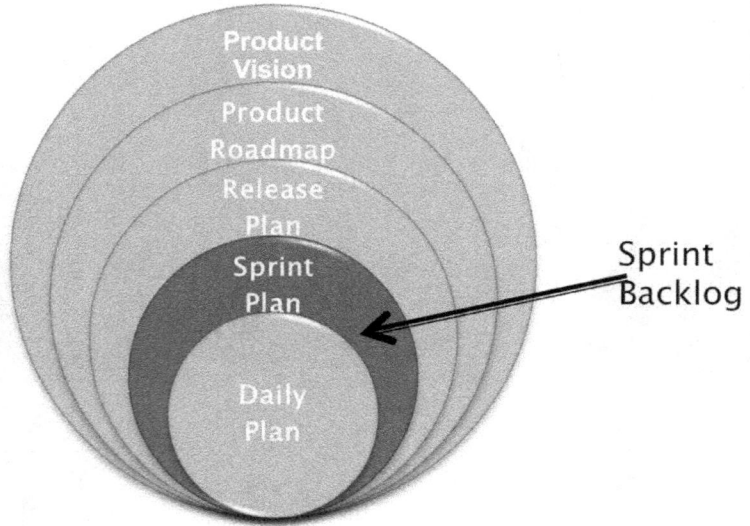

Product Vision

Product Roadmap

Release Plan

Sprint Plan

Daily Plan

Sprint Backlog

CAPE PROJECT MANAGEMENT, INC.

Notes:

Sprint Backlog

- Created at Sprint Planning Meeting and owned by the Development Team
- The Development Team works with the Product owner to select those items that meet the Sprint Goal and will fit in a Sprint
- Based upon story size, priority and velocity
- The Sprint Backlog is emergent and will evolve during the Sprint to meet the Sprint goal

151
Copyrighted material. 2018
CAPE PROJECT MANAGEMENT, INC.

Notes:

Daily Planning

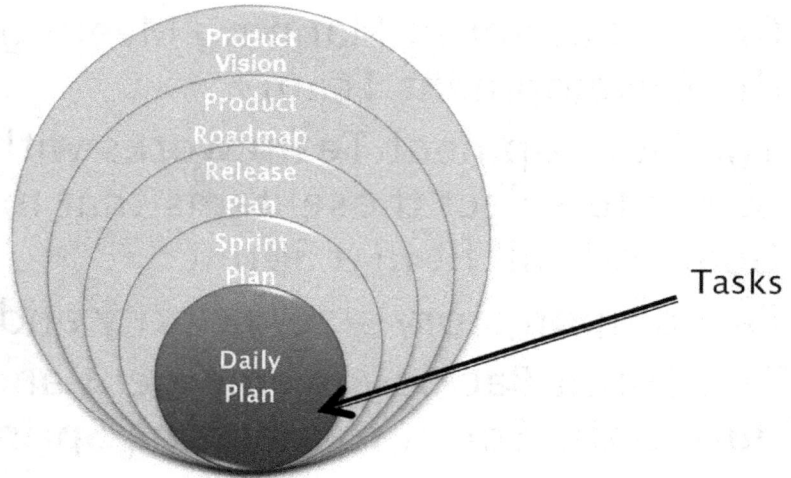

Tasks

CAPE PROJECT
MANAGEMENT, INC.

Notes:

Create User Story Tasks

- ▸ Decompose User Stories into tasks as a team
- ▸ Tasks only need to be defined to cover the next couple of days of work
- ▸ Attempt to size your tasks to take one team member between 4 hours to 2 days to complete
- ▸ Create tasks that result in a deliverable unit of work when completed
- ▸ Don't get caught deep diving into the details of each task

153

CAPE PROJECT
MANAGEMENT, INC.

Notes:

Avoid Mini-Waterfall

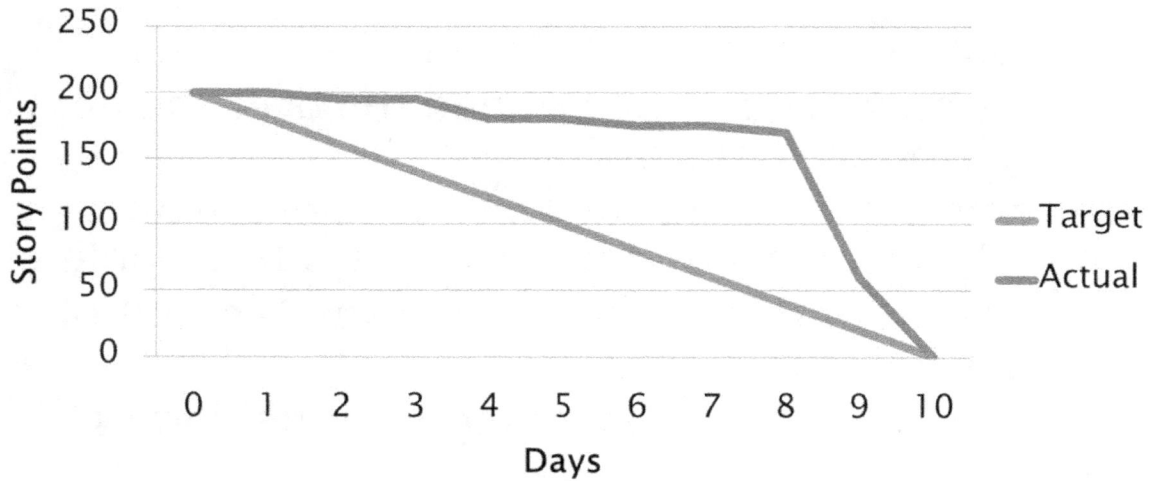

CAPE PROJECT
MANAGEMENT, INC.

Notes:

Team Swarm (or Swarming)

- Working on one or more stories until they are done
- Each story has a "TeamLet"
- Each TeamLet has:
  - A Coordinator who is in charge of the story and stays with it until it is complete
  - A Swarmer(s) brings their expertise to the story. A Swarmer may be on multiple TeamLets.

CAPE PROJECT MANAGEMENT, INC.

Notes:

GROUP EXERCISE

The Sprint Game

156

CAPE PROJECT MANAGEMENT, INC.

Notes:

The Goal

- Accomplish as many of the User Stories you have been given based upon a release plan of 3-5 iterations (Sprints)
- To accomplish this, you will:
  - Prioritize requirements (User Stories)
  - Estimate the size of the User Stories using Story Points
  - Create and execute the plan for the first Sprint
  - Retrospect and plan the next Sprint
  - Track your results and plan using velocity
  - Repeat
- Everyone has to commit!

CAPE PROJECT
MANAGEMENT, INC.

Notes:

Step 1: Backlog Refinement 15 Minutes

- The Product Owner provides User Story cards with business value (Product Backlog)
- The Team estimates complexity of all of the User Stories using T-Shirt Sizes and writes the size on the User Story card.
- The Team orders the Product Backlog by relative priority from largest to smallest using this formula:

X–small	1 Story Point
Small	3 Story Points
Medium	5 Story Points
Large	8 Story Points
X–large	13 Story Points
XX–large	Impossible

$$\frac{\text{Business Value}}{\text{Story Points}} = \text{Relative Priority}$$

CAPE PROJECT MANAGEMENT, INC.

Notes:

Step 2: Create your Sprint Plan
5 Minutes

- Starting with the highest priority Stories on the Product Backlog and select which ones you can complete in a 3 minute Sprint (Sprint Backlog)
- Total the number of Story Points you expect to complete in the Sprint Backlog (Target Velocity)
- Document the Target Velocity on the Sprint Burnup
- Acquire all of the resources you need to execute your Sprint (cards, balloons, etc.)

CAPE PROJECT MANAGEMENT, INC.

Notes:

Step 3: Sprint
3 Minutes

- Execute your User Stories
- Acceptance can occur during the Sprint
- If you finish early, the you can pick additional User Stories from the Product Backlog
- If a story goes badly, you may abandon it...
  - ... if your Product Owner agrees

Copyrighted material. 2018

CAPE PROJECT MANAGEMENT, INC.

Notes:

Step 4: Inspect and Adapt, Plan, Repeat

- ▸ Document Actual Velocity on Burnup chart and re-plan the next Sprint
- ▸ Review and validate estimates
- ▸ Determine new velocity – yesterday's weather
- ▸ Document Target and Actual Velocity on Sprint Burnup chart

Copyrighted material. 2018

CAPE PROJECT
MANAGEMENT, INC.

Notes:

| |
| |
| |
| |
| |
| |
| |
| |

Retrospective

- What did you learn?
- What would you do differently?

 CAPE PROJECT MANAGEMENT, INC.

Notes:

Implementing Agile

CAPE PROJECT
MANAGEMENT. INC.

"We are what we repeatedly do.
Excellence then, is not an act, but a
habit."

~Aristotle

CAPE PROJECT
MANAGEMENT, INC.

Notes:

Scrum is "Easy"

- ‣ Minimal process overhead
- ‣ Less bureaucracy
- ‣ Less documentation
- ‣ Less surprises

CAPE PROJECT
MANAGEMENT, INC.

Notes:

Scrum is Hard

- ‣ Change in processes
- ‣ Change in bureaucracy
- ‣ Change in documentation standards
- ‣ Different kinds of surprises

"When you're finished changing, you're finished."

~Benjamin Franklin

CAPE PROJECT
MANAGEMENT, INC.

Notes:

Key to Success: Change the Paradigm

- ▸ Admit what you did before wasn't working; do something different
- ▸ The only measure of success is delivered software
- ▸ Move to an environment of trust

CAPE PROJECT
MANAGEMENT, INC.

Notes:

Barriers to Implementing Agile

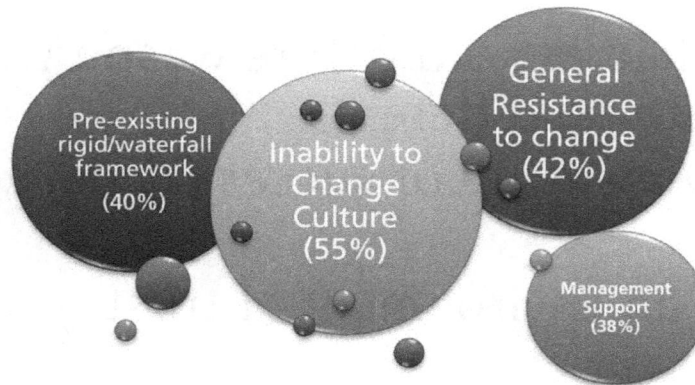

Pre-existing rigid/waterfall framework (40%)

Inability to Change Culture (55%)

General Resistance to change (42%)

Management Support (38%)

... an "inability to change organizational culture" and the "general resistance to change" are the most commonly cited barriers

Source: Version One 2016

CAPE PROJECT MANAGEMENT, INC.

Notes:

Forces of Change

DRIVING Forces

RESISTING Forces

CAPE PROJECT MANAGEMENT, INC.

Notes:

Force Field Analysis

Define the target of change

The center box represents the situation to be moved or changed

Identify which are driving and restraining forces

Analyze the forces to identify which can be changed

Create an action plan to make the changes to the forces

Copyrighted material. 2018

CAPE PROJECT
MANAGEMENT, INC.

Notes:

Activity

Force Field Analysis

CAPE PROJECT
MANAGEMENT, INC.

Notes:

Force Field Analysis

Directions

1. Use the worksheet on the next page.

2. On the center box, write taking change you are anticipating.

3. List all the forces FOR CHANGE in one column, and all the forces AGAINST CHANGE in another column.

4. Rate the strength of these forces and assign a numerical weight, 1 being the weakest, 5 being the strongest.

5. When you add the "strength points" of the forces, you'll see the viability of the proposed change.

The tool can be used to help ensure the success of the proposed change by identifying the strength of the forces against the change.

Forces for Change

Forces against Change

Score

Score

Implementing Agile

Total Score:

Total Score:

GROUP EXERCISE

Agile Jeopardy
http://bit.ly/ScrumGames

CAPE PROJECT
MANAGEMENT, INC.

Thank-You!

CAPE PROJECT
MANAGEMENT, INC.

Cape Project Management, Inc.

Agile Glossary

Note: The following Agile terms and definitions are primarily excerpts from web pages or books. I do not retain the copyright for any of this work. All copyrights reside with their respective authors, web pages or publishers. All other brands or product names used in this Guide are the trade names or registered trademarks of their respective owners.

Agile Contracting	The first lesson we learnt in contracting out agile software development (or anything else for that matter), is to align objectives of the supplier and the customer. It is highly desirable to align supplier success with customer success. The key here is to define the product vision and what must be achieved; foster shared ownership of the goals by treating your supplier as a partner; and consider offering the supplier incentives for meeting key business performance indicators that require partnership with you. http://blog.scrumup.com/2012/11/top-ten-reads-on-agile-contracts.html
Agile Discovery	During the Discovery Phase, designers need to work with the business analyst to capture and define business requirements. This is done by facilitating workshops and interviewing key stakeholders. A lot of sketching, note taking, brainstorming and discussion happen at this stage in order to effectively visualize the early thinking on look and feel, layout and interaction design. It is also worth noting that defining the business model is an evolutionary process. At the end of the discovery and design phases the value proposition needs to map back to real user personas, partnerships, activities, a cost structure, solid business KPIs and have a business mission statement should all clearly defined. https://www.arrkgroup.com/thought-leadership/the-importance-of-getting-the-discovery-phase-right/
Architectural Spike	XP does this while the initial Planning Game is in process. It's not an iteration - it might be longer or shorter, we don't know yet. What it's about is exploring solution elements that seem relevant to the as yet limited knowledge we have about the problem domain, choosing a System Metaphor, putting enough of our build, dbms, and source control tools in place to be able to begin controlled work, and then proceeding until we have something that runs and can be iterated. https://www.linkedin.com/pulse/agile-architecting-practice-architecture-spike-erik-philippus
Active Listening	Active listening is a communication technique used in counselling, training and conflict resolution, which requires the listener to feed back what they hear to the speaker, by way of re-stating or paraphrasing what they have heard in their own words, to confirm what they have heard and moreover, to confirm the understanding of both parties https://en.wikipedia.org/wiki/Active_listening
Adaptive Leadership	Adaptive leadership focuses on team management, from building self-organizing teams to developing a servant leadership style. It is both more difficult, and ultimately more rewarding than managing tasks. In an agile enterprise the people take care of the tasks and the leader engages the people. The facilitative leader works on things like building self-organizing teams, a trusting and respectful environment, collaboration, participatory decision making, and developing appropriate empowerment guidelines (for an excellent discussion of empowerment, see Chapters 6 & 7 in (Appelo, 2011)). https://www.infoq.com/news/2011/02/highsmith-adaptive-leadership Appelo, J. (2011). Management 3.0: Leading Agile Developers, Developing Agile Leaders. Upper Saddle River, JN: Addison-Wesley.

Adaptability, Three Components	Adaptability has three components—**product, process, and people.** You need to have a gung-ho agile team with the right attitude about change. You need processes and practices that allow the team to adapt to circumstances. And you *need* high quality code with automated tests. You can have pristine code and a non-agile team and change will be difficult. All three are required to have an agile, adaptable environment. http://searchsoftwarequality.techtarget.com/feature/Adaptation-in-project-management-through-agile
Affinity Estimating	A facilitated process where team members for sequence the product backlog from smallest to largest user story, then the rest of the team validates and finally the user stories are group by a sizing method such as t-shirt size or Fibonacci sequence. http://www.gettingagile.com/2008/07/04/affinity-estimating-a-how-to/
Agile Product Lifecycle	**The Five Agile plans:** 1 – Product Vision: Yearly By Product Owner 2 – Product Roadmap: Bi-Yearly By The Product Owner 3 – Release Plan: Quarterly By The Product Owner And Teams 4 – Iteration Plan: Bi-Weekly By The Teams 5 – Daily Plan (Scrum): Daily By Individual (EXAM TIP: different authors reference different time periods for updating these plans) https://www.ramantech.com/implementing-the-5-levels-of-agile-planning/ http://www.romanpichler.com/blog/
Agile Modeling (AM)	AM is a collection of values, principles, and practices for modeling software that can be applied on a software development project in an effective and light-weight manner. Values: 1. Communication 2. Simplicity 3. Feedback 4. Courage 5. Humility Principles/Practices: • Active Stakeholder Participation. Stakeholders should provide information in a timely manner, make decisions in a timely manner, and be as actively involved in the development process through the use of inclusive tools and techniques. • Architecture Envisioning. At the beginning of an agile project you will need to do some initial, high-level architectural modeling to identify a viable technical strategy for your solution. • Document Continuously. Write deliverable documentation throughout the lifecycle in parallel to the creation of the rest of the solution. • Document Late. Write deliverable documentation as late as possible, avoiding speculative ideas that are likely to change in favor of stable information. • Executable Specifications. Specify requirements in the form of executable "customer tests", and your design as executable developer tests, instead of non-executable "static" documentation. • Iteration Modeling. At the beginning of each iteration you will do a bit of modeling as

	part of your iteration planning activities. • Just Barely Good Enough (JBGE) artifacts. A model or document needs to be sufficient for the situation at hand and no more. • Look Ahead Modeling. Sometimes requirements that are nearing the top of your priority stack are fairly complex, motivating you to invest some effort to explore them before they're popped off the top of the work item stack so as to reduce overall risk. • Model Storming. Throughout an iteration you will model storm on a just-in-time (JIT) basis for a few minutes to explore the details behind a requirement or to think through a design issue. • Multiple Models. Each type of model has its strengths and weaknesses. An effective developer will need a range of models in their intellectual toolkit enabling them to apply the right model in the most appropriate manner for the situation at hand. • Prioritized Requirements. Agile teams implement requirements in priority order, as defined by their stakeholders, so as to provide the greatest return on investment (ROI) possible. • Requirements Envisioning. At the beginning of an agile project you will need to invest some time to identify the scope of the project and to create the initial prioritized stack of requirements. • Single Source Information. Strive to capture information in one place and one place only. • Test-Driven Design (TDD). Write a single test, either at the requirements or design level, and then just enough code to fulfill that test. TDD is a JIT approach to detailed requirements specification and a confirmatory approach to testing. http://www.agilemodeling.com/
Agile Scaling Model	Agile scaling factors are: 1. Team size 2. Geographical distribution 3. Regulatory compliance 4. Domain complexity 5. Organizational distribution 6. Technical complexity 7. Organizational complexity 8. Enterprise discipline https://www.agilealliance.org/wp-content/uploads/2016/01/Agile-Scaling-Model.pdf
Agile Triangle	The three dimensions critical to Agile performance measurement : • Value, • Quality • Constraints (cost, schedule, scope). Also, simplified to be: • Value • Technical debt • Cost http://jimhighsmith.com/beyond-scope-schedule-and-cost-the-agile-triangle/
APM Delivery Framework	However, if the business objective is reliable innovation, then the process framework must be organic, flexible, and easy to adapt. The APM process framework supports this second business objective through the five phases of:

	• Envision • Speculate • Explore • Adapt • Close http://www.informit.com/articles/article.aspx?p=174660&seqNum=4
Asynchronous Builds	When you use the integration script discussed earlier, you're using synchronous integration—you're confirming that the build and tests succeed before moving on to your next task. If the build is too slow, synchronous integration becomes untenable. (For me, 20 or 30 minutes is too slow.) In this case, you can use asynchronous integration instead. Rather than waiting for the build to complete, start your next task immediately after starting the build, without waiting for the build and tests to succeed. The biggest problem with asynchronous integration is that it tends to result in broken builds. If you check in code that doesn't work, you have to interrupt what you're doing when the build breaks half an hour or an hour later. If anyone else checked out that code in the meantime, their build won't work either. If the pair that broke the build has gone home or to lunch, someone else has to clean up the mess. In practice, the desire to keep working on the task at hand often overrides the need to fix the build. If you have a very slow build, asynchronous integration may be your only option. If you must use this, a continuous integration server is the best way to do so. It will keep track of what to build and automatically notify you when the build has finished. http://jamesshore.com/Agile-Book/continuous_integration.html
Backlog Grooming/ Refinement	Product Backlog refinement is the act of adding detail, estimates, and order to items in the Product Backlog. This is an ongoing process in which the Product Owner and the Development Team collaborate on the details of Product Backlog items. During Product Backlog refinement, items are reviewed and revised. The Scrum Team decides how and when refinement is done. Refinement usually consumes no more than 10% of the capacity of the Development Team. However, Product Backlog items can be updated at any time by the Product Owner or at the Product Owner's discretion. http://www.scrumguides.org/scrum-guide.html
Brainstorming	A brainstorming session is a tool to generate ideas from a selected audience to solve a problem or stimulate creativity. These meetings are used for solving a process problem, inventing new products or product innovation, solving inter-group communication problems, improving customer service, budgeting exercises, project scheduling, etc. Using these tools can help discussion facilitators and Project Managers with alternative approaches for creative idea generation meetings (aka Brainstorming Sessions). They are particularly useful when previous meetings have gone afoul, are not as effective as they could be or productivity during these exercises is less than it should be. http://www.projectconnections.com/templates/detail/brainstorming-meeting-techniques.html
Burn Down Charts	

X-Axis	The project/iteration timeline

	Y-Axis	The work that needs to be completed for the project. The time estimates for the work remaining will be represented by this axis.
	Project Start Point	This is the farthest point to the left of the chart and occurs at day 0 of the project/iteration.
	Project End Point	This is the point that is farthest to the right of the chart and occurs on the predicted last day of the project/iteration
	Ideal Work Remaining Line	This is a straight line that connects the start point to the end point. At the start point, the ideal line shows the sum of the estimates for all the tasks (work) that needs to be completed. At the end point, the ideal line intercepts the x-axis showing that there is no work left to be completed.
	Actual Work Remaining Line	This shows the actual work remaining. At the start point, the actual work remaining is the same as the ideal work remaining but as time progresses; the actual work line fluctuates above and below the ideal line depending on how effective the team is. In general, a new point is added to this line each day of the project. Each day, the sum of the time estimates for work that was recently completed is subtracted from the last point in the line to determine the next point.
	Actual Work Line is above the Ideal Work Line	If the actual work line is above the ideal work line, it means that there is more work left than originally predicted and the project is behind schedule.
	Actual Work Line is below the Ideal Work Line	If the actual work line is below the ideal work line, it means that there is less work left than originally predicted and the project is ahead of schedule.

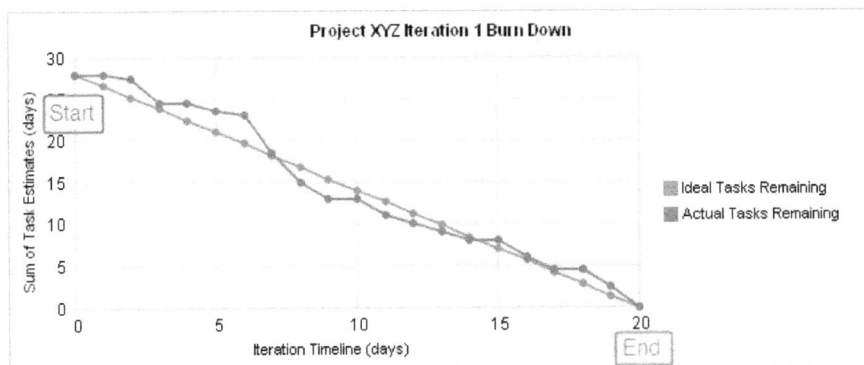

Project XYZ Iteration 1 Burn Down

http://en.wikipedia.org/wiki/Burn_down_chart

Burndown Bar Charts	The typical Scrum release burndown bar chart shows a single value--the net change in the amount of work remaining. • As tasks are completed, the top of the bar is lowered. • When tasks are added to the original set, the bottom of the bar is lowered. • When tasks are removed from the original set, the bottom of the bar is raised. • When the amount of work involved in a task changes, the top of the bar moves up or down.

Prioritized Task List
Action Item Burndown

https://www.mountaingoatsoftware.com/agile/scrum/scrum-tools/release-burndown/alternative

Burn Up Charts	The amount of accepted work (that work which has been completed, tested, and met acceptance criteria) is graphed in a burnup chart. The amount of work in an accepted state starts at 0 and continues to grow until it reaches 100% accepted at the end of the Iteration.

Iteration Burnup
The Chuck Norris - Hard 149 (2009.1) (03/02/09 - 03/06/09)

http://www.clariostechnology.com/productivity/blog/whatisaburnupchart

Chartering in Agile	1. **Vision:** The vision defines the "Why" of the project. This is the higher purpose, or the reason for the project's existence. 2. **Mission:** This is the "What" of the project and it states what will be done in the project to achieve its higher purpose. 3. **Success Criteria:** The success criteria are management tests that describe effects outside of the solution itself. https://www.infoq.com/news/2010/05/agile-project-charter
Collaboration	Collaboration is the basis for bringing together the knowledge, experience and skills of multiple team members to contribute to the development of a new product more effectively than individual team members performing their narrow tasks in support of product development. As such collaboration is the basis for concepts such as concurrent engineering or integrated product development.

	Effective collaboration requires actions on multiple fronts: • Early involvement and the availability of resources to effectively collaborate • A culture that encourages teamwork, cooperation and collaboration • Effective teamwork and team member cooperation • Defined team member responsibilities based on collaboration • A defined product development process based on early sharing of information and collaborating • Collocation or virtual collocation • Collaboration technology (EXAM TIP: There was specifically a question about the difference between coordination and collaboration) http://www.npd-solutions.com/collaboration.html
Collaboration Games	Agile games are activities focused on teaching, demonstrating, and improving Agile and organizational effectiveness using game theory. Using games, we can model complex or time-consuming processes and systems, examine why they work (or don't work), look for improvements, and teach others how to benefit from them. Games can model just the core of a process or model, leaving out unimportant factors. They can involve collaboration, brainstorming, comparing variants, and of course retrospectives. http://tastycupcakes.org/
Collective Code Ownership	The way this works is for each developer to create unit tests for their code as it is developed. All code that is released into the source code repository includes unit tests that run at 100%. Code that is added, bugs as they are fixed, and old functionality as it is changed will be covered by automated testing. Now you can rely on the test suite to watch dog your entire code repository. Before any code is released it must pass the entire test suite at 100%. Once this is in place anyone can make a change to any method of any class and release it to the code repository as needed. When combined with frequent integration developers rarely even notice a class has been extended or repaired. http://www.extremeprogramming.org/rules/collective.html
Collocated And Or Distributed Teams	Collocated Agile is a model in which projects execute the Agile Methodology with teams located in a single room. The methodology requires that the complete team be in close proximity to each other to improve coordination between the members. Collocated Agile teams have proven that the real power of project success lies not in administration, but in the acumen, chemistry, loyalty, and dedication between the collocated teams. Distributed Agile, as the name implies, is a model in which projects execute an Agile Methodology with teams that are distributed across multiple geographies. http://www.continuousagile.com/unblock/team_options.html
Compliance	Agile allows faster time to market by deploying working code more quickly. Application of user stories and iterations in a regulated industry can help to keep the team focused for large projects. Compliance regulations or legislation can sometimes change quickly. The opportunity to more easily cater for change gives the business stakeholders value earlier. As requirements change there is an avoidance of waste, due to not designing and detailing all requirements upfront. A successful agile compliance project requires the right people, the right team culture and the right amount of regulation / industry knowledge in the team.

	A tool may assist in enabling the appropriate levels of communication and reporting – so that the status of the project can be easily visible. Various considerations for agile projects in a regulated industry have been discussed here. Although there may be challenges, for a team with the right focus and frame of mind it is a matter of "where there is a will there is a way". Some areas may need to be adapted – however that merely seems to be in line with the whole essence and concept of what agile is all about – agility and responsiveness. https://www.planittesting.com/us/Insights/2012/Agile-and-Regulatory-Compliance
Conflict Resolution	Principles to Remember Do: Set up conflict management procedures before a conflict arises Intervene early when a fight erupts between team members Get the team working together again as soon as possible Don't: Assume your team agrees on its shared purpose, values, or vision Let conflicts fester or go unattended Move on without first talking about the conflict as a team https://hbr.org/2010/06/get-your-team-to-stop-fighting.html
Conflict Levels	**Level 1: Problem to Solve** We all know what conflict at level 1 feels like. Everyday frustrations and aggravations make up this level, and we experience conflicts as they rise and fall and come and go. At this level, people have different opinions, misunderstanding may have happened, conflicting goals or values may exist, and team members likely feel anxious about the conflict in the air. When in level 1, the team remains focused on determining what's awry and how to fix it. Information flows freely, and collaboration is alive. Team members use words that are clear, specific, and factual. The language abides in the here and now, not in talking about the past. Team members check in with one another if they think a miscommunication has just happened. You will probably notice that team members seem optimistic, moving through the conflict. It's not comfortable, but it's not emotionally charged, either. Think of level 1 as the level of constructive disagreement that characterizes high-performing teams. **Level 2: Disagreement** At level 2, self-protection becomes as important as solving the problem. Team members distance themselves from one another to ensure they come out OK in the end or to establish a position for compromise they assume will come. They may talk offline with other team members to test strategies or seek advice and support. At this level, good-natured joking moves toward the half-joking barb. Nastiness gets a sugarcoating but still comes across as bitter. Yet, people aren't hostile, just wary. Their language reflects this as their words move from the specific to the general. Fortifying their walls, they don't share all they know about the issues. Facts play second fiddle to interpretations and create confusion about what's really happening. **Level 3: Contest** At level 3, the aim is to win. A compounding effect occurs as prior conflicts and problems remain unresolved. Often, multiple issues cluster into larger issues or create a "cause." Factions emerge in this fertile ground from which misunderstandings and power politics arise. In an agile team, this may happen subtly, because a hallmark of working agile is the feeling that we are all in this together. But it does happen. People begin to align themselves with one side or the other. Emotions become tools used to

"win" supporters for one's position. Problems and people become synonymous, opening people up to attack. As team members pay attention to building their cases, their language becomes dis-torted. They make overgeneralizations: "He always forgets to check in his code" or "You never listen to what I have to say." They talk about the other side in presumptions: "I know what they think, but they are ignoring the real issue." Views of themselves as benevolent and others as tarnished become magnified: "I am always the one to compromise for the good of the team" or "I have everyone's best interest at heart" or "They are intentionally ignoring what the customer is really saying." Discussion becomes either/or and blaming flourishes. In this combative environment, talk of peace may meet resistance. People may not be ready to move beyond blaming.

Level 4: Crusade
At level 4, resolving the situation isn't good enough. Team members believe the people on the "other side" of the issues will not change. They may believe the only option is to remove the others from the team or get removed from the team themselves. Factions become entrenched and can even solidify into a pseudo-organizational structure within the team. Identifying with a faction can overshadow identifying with the team as a whole so the team's identity gets trounced. People and positions are seen as one, opening up people to attack for their affiliations rather than their ideas. These attacks come in the form of language rife with ideology and principles, which becomes the focus of conversation, rather than specific issues and facts. The overall attitude is righteous and punitive.

Level 5: World War
"Destroy!" rings out the battle cry at level 5. It's not enough that one wins; others must lose. "We must make sure this horrible situation does not happen again!" Only one option at level 5 exists: to separate the combatants (aka team members) so that they don't hurt one another. No constructive outcome can be had.

http://agile.dzone.com/articles/agile-managing-conflict

Conflict Types	**#1 - Lack of Role Clarity** The project manager is responsible for assigning tasks to each project team member. In addition, they often assume that team members understand what is being asked of them. This assumption can be incorrect, leading to team members being unclear on what needs to be accomplished. A good project manager takes the time to explain the tasks, their expectations and timeframes around completion. **#2 - Difference in Prioritizing Tasks** Just because the project manager thinks the task is a milestone, the team member completing the task may not. Team members may be working simultaneously on multiple projects and cannot differentiate the priority of one project's tasks from another. The project manager should try to explain the importance of the overall project to the **#3 - Working in Silos** Often, project team members work independently. They may work remotely or in a different location from other project team members. Conflict arises when team members are not aware of what others are doing and are not communicating with one another. The project manager needs to bring the team together to discuss project status and barriers to getting the project completed promptly. If team members working in silos can envision how they are a part of the bigger picture, they will be more motivated and feel like a part of the team. **#4 - Lack of Communication** Project managers must foster a clear line of communication between project team members. In order to minimize duplication of efforts, the project manager should communicate expectations

	to all team members. The project manager needs to be easily accessible to project team members at all times during the project. If team members cannot reach their project manager or other team members, they may spin in circles needlessly. **# 5 - Waiting on Completion of Task Dependencies** Some tasks cannot be started until other tasks are completed. Team members need to understand the impact of their role on others. For example if one team member is responsible for ordering equipment and another for installing the equipment, one task is dependent on the other. Conflict can occur if the first team member is delayed in completing their tasks. http://www.brighthub.com/office/project-management/articles/95971.aspx
Container	A container is a closed space where things can get done, regardless of the overall complexity of the problem. In the case of Scrum, a container is a Sprint, an iteration. We put people with all the skills needed to solve the problem in the container. We put the highest value problems to be solved into the container. Then we protect the container from any outside disturbances while the people attempt to bring the problem to a solution. We control the container by time-boxing the length of time that we allow the problem to be worked on. We let the people select problems of a size that can be brought to fruition during the time-box. At the end of the time-box, we open the container and inspect the results. We then reset the container (adaptation) for the next time-box. By frequently replanning and shifting our work, we are able to optimize value. http://kenschwaber.wordpress.com/2010/06/10/waterfall-leankanban-and-scrum-2/
Continuous Integration	Continuous Integration (CI) involves producing a clean build of the system several times per day. Agile teams typically configure CI to include automated compilation, unit test execution, and source control integration. Sometimes CI also includes automatically running automated acceptance tests. **Continuous Integration Technique, Tools, and Policy** There are several specific practices that CI seems to require to work well. On his site, Martin Fowler provides a long, detailed description of what Continuous Integration is and how to make it work. One popular CI rule states that programmers never leave anything unintegrated at the end of the day. The build should never spend the night in a broken state. This imposes some task planning discipline on programming teams. Furthermore, if the team's rule is that whoever breaks the build at check-in has to fix it again, there is a natural incentive to check code in frequently during the day. **Benefits of Continuous Integration** When CI works well, it helps the code stay robust enough that customers and other stakeholders can play with the code whenever they like. This speeds the flow of development work overall; as Fowler points out, it has a very different feel to it. It also encourages more feedback between programmers and customers, which helps the team get things right before iteration deadlines. Like refactoring, continuous integration works well if you have an exhaustive suite of automated unit tests that ensure that you are not committing buggy code. http://www.versionone.com/Agile101/Continuous_Integration.asp http://www.martinfowler.com/articles/continuousIntegration.html
Control Limits	Control limits, also known as natural process limits, are horizontal lines drawn on a statistical process control chart, usually at a distance of ±3 standard deviations of the plotted statistic from the statistic's mean. Used in Agile Control charts

	https://en.wikipedia.org/wiki/Control_limits
Cumulative Flow Diagrams	The Cumulative Flow diagram is very similar to a Burnup Chart. It shows how much of our work (i.e., the effort associated with User Stories) is in different states, such as Completed or In Progress. The Total and Completed curves shows the Release scope and Burnup of completed work, while the In Progress curve shows how much work is associated with Stories currently in development. The primary difference from the standard Burnup and Cumulative Flow diagrams is that the latter shows how much of the work is currently in progress. https://www.cprime.com/resource/templates/cumulative-flow-diagram-burnup-chart/
Customer-Valued Prioritization	Agile development is about the frequent delivery of high-value, working software to the customer/user community. Doing so requires the prioritization of user stories and the continuous monitoring of the prioritized story backlog. The primary driver for prioritization is customer value. However, it is insufficient to simply say that the highest-value stories are the highest priority. Product owners must also factor in the cost of development. An extremely valuable feature quickly loses its luster when it is also extremely costly to implement. Additionally, there are other secondary drivers such as risk and uncertainty. These should be resolved early. There may also be experimental stories that are worth developing early to find out whether customers see value in further development along those lines. There may be other prioritization drivers, but business value should always be foremost. 1. Complete the high-value, high-risk stories first if the cost is justified. 2. Complete the high-value, low-risk stories next if the cost is justified. 3. Complete the lower-value, low-risk stories next. 4. Avoid low-value, high-risk stories. http://www.scribd.com/doc/111905434/Agile-Analytics-a-Value-Driven-Approach-Ken-Collier
Cycle Time	Cycle time for software development is measured in the number of days needed between feature specification and production delivery. This is called: Software In Process (SIP). A shorter cycle indicates a healthier project. A Lean project that deploys to production every 2-weeks has a SIP of 10 working days. Some Lean projects even deploy nightly. https://jaymeedwards.com/2012/04/09/cycle-time-the-important-statistic-you-probably-arent-measuring/

Daily Scrums/Daily Stand-Ups	Time-boxed 15 minutes meetings whose purpose is the providing of a concise team status. Scrum has each team member asked by the ScrumMaster the following questions: • What did you do yesterday? • What are you doing today? • Are there any impediments that need resolution? (EXAM TIP: I had a couple of questions relating to the purpose of this meeting) http://www.mountaingoatsoftware.com/scrum/daily-scrum
Declaration Of Interdependence	The **PM Declaration of interdependence** is a set of six management principles initially intended for project managers of Agile Software Development projects We are a community of project leaders that are highly successful at delivering results. To achieve these results: • We **increase return on investment** by making continuous flow of value our focus. • We **deliver reliable results** by engaging customers in frequent interactions and shared ownership. • We **expect uncertainty** and manage for it through iterations, anticipation, and adaptation. • We **unleash creativity and innovation** by recognizing that individuals are the ultimate source of value, and creating an environment where they can make a difference. • We **boost performance** through group accountability for results and shared responsibility for team effectiveness. • We **improve effectiveness and reliability** through situationally specific strategies, processes and practices. http://pmdoi.org/
D.E.E.P.	• **Detailed Appropriately.** User stories on the product backlog that will be done soon need to be sufficiently well understood that they can be completed in the coming sprint. Stories that will not be developed for awhile should be described with less detail. • **Estimated.** The product backlog is more than a list of all work to be done; it is also a useful planning tool. Because items further down the backlog are not as well understood (yet), the estimates associated with them will be less precise than estimates given items at the top. • **Emergent.** A product backlog is not static. It will change over time. As more is learned, user stories on the product backlog will be added, removed, or reprioritized. • **Prioritized.** The product backlog should be sorted with the most valuable items at the top and the least valuable at the bottom. By always working in priority order, the team is able to maximize the value of the product or system being developed. http://www.mountaingoatsoftware.com/blog/make-the-product-backlog-deep
Definition Of Done	Definition of Done (DoD) is a simple list of activities (writing code, coding comments, unit testing, integration testing, release notes, design documents, etc.) that add verifiable/demonstrable value to the product. Focusing on value-added steps allows the team to focus on what must be completed in order to build software while eliminating wasteful activities that only complicate software development efforts. Note – the DoD is defined by the Product Owner and committed to by the team. http://www.scrumalliance.org/articles/105-what-is-definition-of-done-dod

Defect Rate	The defect detection rate is the amount of defects detected per sprint. Assuming that developers produce defects at a more or less constant rate, it is correlated with the velocity; the more story points are delivered, the more defects should be found and fixed as well. Teams tend to be pretty consistent in the quality of the software they deliver, so a drop in velocity combined with a rise in the defect detection rate should trigger the alarm. Something's cooking and you need to find out what it is. My personal opinion is that a lower defect detection rate isn't necessarily better than a higher one. A defect more found in one of the development and test environments is a defect less that makes it into production. From that perspective, you could support the statement, the more defects the better. http://theagileprojectmanager.blogspot.com/2013/05/agile-metrics.html
Discounted Pay-Back Period	A capital budgeting procedure used to determine the profitability of a project. In contrast to an NPV analysis, which provides the overall value of a project, a discounted payback period gives the number of years it takes to break even from undertaking the initial expenditure. Future cash flows are considered are discounted to time "zero." This procedure is similar to a payback period; however, the discounted payback period measures how long it take for the initial cash outflow to be paid back, including the time value of money. http://accountingexplained.com/managerial/capital-budgeting/discounted-payback-period
Dreyfus Model of Skill Acquisition	**1. Novice** • "rigid adherence to taught rules or plans" • no exercise of "discretionary judgment" **2. Advanced beginner** • limited "situational perception" • all aspects of work treated separately with equal importance **3. Competent** • "coping with crowdedness" (multiple activities, accumulation of information) • some perception of actions in relation to goals • deliberate planning • formulates routines **4. Proficient** • holistic view of situation • prioritizes importance of aspects • "perceives deviations from the normal pattern" • employs maxims for guidance, with meanings that adapt to the situation at hand **5. Expert** • transcends reliance on rules, guidelines, and maxims • "intuitive grasp of situations based on deep, tacit understanding" • has "vision of what is possible" • uses "analytical approaches" in new situations or in case of problems https://en.wikipedia.org/wiki/Dreyfus_model_of_skill_acquisition
DRY	Software must be written expecting for future change. Principles like **D**on't **R**epeat **Y**ourself (DRY) are used to facilitate this. In agile development, changes to the software specifications are welcome even in late stages of development. As clients get more hands-on time with iterative builds of the software, they may be able to better communicate their needs. Also, changes in the market or company structure might dictate changes in the software specifications. Agile development is designed to accommodate these late changes.

	https://code.tutsplus.com/tutorials/3-key-software-principles-you-must-understand--net-25161
Earned Value Management (EVM)	Agile EVM is now all about executing the project and tracking the accumulated EV according to the simple earning rule. Because Agile EVM has been evolving for many years the following practices are well-established: • EV is accumulated at fixed time intervals (i.e. Timebox, Iteration or Sprint) of 1–4 weeks; • PV and EV is graphically tracked & extrapolated as remaining value in a Release Burndown Chart as shown in figure 6; • Rather than a S-shaped curve the PV in Agile EVM is a straight line because an Agile project has no distinct phases and corresponding variances in the rate of value delivery; • The EV in Story Points done in one fixed time interval is known as the Velocity of a team; • In Agile scope change is embraced and the amount of added (removed) scope in Story Points is added (removed) to the Velocity or Scope Floor. The latter is shown in Figure 6 where several scope increases have lowered the Scope Floor below the x-axis. The advantage of using a Scope Floor is that any scope changes can easily be separated from Velocity variances; • The intersection between the Remaining Value and Scope Floor lines indicates the expected release date and the corresponding Remaining Budget. (EXAM TIP –There was a question on the exam that asked when was the best time to measure EVM) http://en.wikipedia.org/wiki/Earned_value_management#Agile_EVM
Emotional Intelligence	**Self Awareness** Awareness of one's own feelings and the ability to recognize and manage these feelings in a way which one feels that one can control. This factor includes a degree of self-belief in one's ability to manage one's emotions and to control their impact in a work environment. **Emotional Resilience** The ability to perform consistently in a range of situations under pressure and to adapt one's behavior appropriately. The facility to balance the needs and concerns of the individuals involved. The ability to retain focus on a course of action or need for results in the face of personal challenge or criticism. **Motivation** The drive and energy to achieve clear results and make an impact: and to balance both short and long term goals with an ability to pursue demanding goals in the face of rejection or questioning. **Interpersonal Sensitivity** The facility to be aware of, and take account of, the needs and perceptions of others when arriving at decisions and proposing solutions to problems and challenges. The ability to build from this awareness and achieve 'buy-in' to decisions and ideas for action. **Influence** The ability to persuade others to change a viewpoint based on the understanding of the position and the recognition of the need to listen to this perspective and provide a rationale for change. **Intuitiveness** The ability to arrive at clear decisions and drive their implementation when presented with incomplete or ambiguous information using both rational and 'emotional' or insightful perceptions of key issues and implications. **Conscientiousness** The ability to display clear commitment to a course of action in the face of challenge and to match 'words and deeds'; in encouraging others to support the chosen direction. The personal

	commitment to pursuing an ethical solution to a difficult business issue or problem. http://businessagile.blogspot.com/2005/02/emotional-intelligence-key-element-of.html
Empirical Process Control	**Empirical Process Control** The word "empirical" denotes information gained by means of observation, experience, or experiment. The empirical process control constitutes a continuous cycle of inspecting the process for correct operation and results and adapting the process as needed. There are three legs that hold up every implementation of empirical process control: **transparency, inspection, and adaptation**. The first leg is transparency. It means that those aspects of the process that affect the outcome must be visible to those controlling the process. The second leg is inspection. The various aspects of the process must be inspected frequently enough that unacceptable variances in the process can be detected. The third leg of empirical process control is adaptation. The process or the material being processed must be adjusted if one or more aspects of the process are outside acceptable limits and the resulting product will be unacceptable. http://www.scrumstudy.com/scrum-empirical-process-control.asp
Escaped Defects	Escaped Defects are those defects reported by the Customer that have escaped all software quality processes are represented in this metric. Escaping defects should then be treated as ranked backlog work items, along with other project work items. They should be prioritized high enough to resolve them within the next sprint or two and not accumulate a growing backlog. Watch the defect backlog as part of the project metrics. A growing defect backlog is a key indicator that the team is taking on more new work than it can handle. It may also be a key indicator that the team is operating as a "mini-waterfall" project, rather than a agile project, requiring more collaboration between Dev and Quality Engineers and early testing. Drop the number of new items the team works on until the escaping defects are well managed or eliminated. http://qablog.practitest.com/process-quality-feedback-and-escaping-defects/
Exploratory Testing	Exploratory Testing is a technique for finding surprising defects. Testers use their training, experience, and intuition to form hypotheses about where defects are likely to be lurking in the software, then they use a fast feedback loop to iteratively generate, execute, and refine test plans that expose those defects. It appears similar to ad-hoc testing to an untrained observer, but it's far more rigorous. Some teams use exploratory testing to check the quality of their software. After a story's been coded, the testers do some exploratory testing, the team fixes bugs, and repeat. Once the testers don't find any more bugs, the story is done. (EXAM TIP: This is also referred to as testing performed between Done and Done-Done) http://jamesshore.com/Blog/Alternatives-to-Acceptance-Testing.html
Extreme Programming	Extreme Programming (or XP) is a set of values, principles and practices for rapidly developing high-quality software that provides the highest value for the customer in the fastest way possible. XP is extreme in the sense that it takes 12 well-known software development "best practices" to their logical extremes. The 12 core practices of XP are:

1) **The Planning Game**: Business and development cooperate to produce the maximum business value as rapidly as possible. The planning game happens at various scales, but the basic rules are always the same:
   a) Business comes up with a list of desired features for the system. Each feature is written out as a **User Story**, which gives the feature a name, and describes in broad strokes what is required. User stories are typically written on 4x6 cards.
   b) Development estimates how much effort each story will take, and how much effort the team can produce in a given time interval (the iteration).
   c) Business then decides which stories to implement in what order, as well as when and how often to produce a production releases of the system.
2) **Small Releases**: Start with the smallest useful feature set. Release early and often, adding a few features each time.
3) **System Metaphor**: Each project has an organizing metaphor, which provides an easy to remember naming convention.
4) **Simple Design**: Always use the simplest possible design that gets the job done. The requirements will change tomorrow, so only do what's needed to meet today's requirements.
5) **Continuous Testing**: Before programmers add a feature, they write a test for it. When the suite runs, the job is done. Tests in XP come in two basic flavors.
   a) **Unit Tests** are automated tests written by the developers to test functionality as they write it. Each unit test typically tests only a single class, or a small cluster of classes. Unit tests are typically written using a unit testing framework, such as JUnit.
   b) **Acceptance Tests** (also known as **Functional Tests**) are specified by the customer to test that the overall system is functioning as specified. Acceptance tests typically test the entire system, or some large chunk of it. When all the acceptance tests pass for a given user story, that story is considered complete. At the very least, an acceptance test could consist of a script of user interface actions and expected results that a human can run. Ideally acceptance tests should be automated, either using the unit testing framework, or a separate acceptance testing framework.
6) **Refactoring**: Refactor out any duplicate code generated in a coding session. You can do this with confidence that you didn't break anything because you have the tests.
7) **Pair Programming**: All production code is written by two programmers sitting at one machine. Essentially, all code is reviewed as it is written.
8) **Collective Code Ownership**: No single person "owns" a module. Any developer is expect to be able to work on any part of the codebase at any time.
9) **Continuous Integration**: All changes are integrated into the codebase at least daily. The tests have to run 100% both before and after integration.
10) **40-Hour Work Week**: Programmers go home on time. In crunch mode, up to one week of overtime is allowed. But multiple consecutive weeks of overtime are treated as a sign that something is very wrong with the process.
11) **On-site Customer**: Development team has continuous access to a real live customer, that is, someone who will actually be using the system. For commercial software with lots of customers, a customer proxy (usually the product manager) is used instead.
12) **Coding Standards**: Everyone codes to the same standards. Ideally, you shouldn't be able to tell by looking at it who on the team has touched a specific piece of code.

(EXAM TIP: Lots of questions about Lean, XP and Scrum principles)

http://www.jera.com/techinfo/xpfaq.html

Feature	During detailed planning, agile development favors a feature breakdown structure (FBS)

Breakdown Structure (FBS)	approach instead of the work breakdown structure (WBS) used in waterfall development approaches. Feature breakdown structures are advantageous for a few reasons: 1. They allow communication between the customer and the development team in terms both can understand. 2. They allow the customer to prioritize the team's work based on business value. 3. They allow tracking of work against the actual business value produced. It is acceptable to start out with features that are large and then break them out into smaller features over time. This allows the customer to keep from diving in to too much detail until that detail is needed to help facilitate actual design and delivery. http://www.versionone.com/agile-101/agile-feature-estimation/
Fishbone diagram	The fishbone diagram identifies many possible causes for an effect or problem. It can be used to structure a brainstorming session. It immediately sorts ideas into useful categories. http://asq.org/learn-about-quality/cause-analysis-tools/overview/fishbone.html
The Five Whys	A technique for identifying the exact root cause of a problem to determine the appropriate solution. Made popular in the 70s by the Toyota Production System, the 5 whys is a flexible problem solving technique. http://businessanalystlearnings.com/ba-techniques/2013/2/5/root-cause-analysis-the-5-whys-technique
Fractional Assignments	All of the team members should sit with the team full-time and give the project their complete attention. This particularly applies to customers, who are often surprised at the level of involvement XP requires of them. Some organizations like to assign people to multiple projects simultaneously. This fractional assignment is particularly common in matrix-managed organizations. (If team members have two managers, one for their project and one for their function, you're probably in a matrixed organization.) Fractional assignment is dreadfully counterproductive. If your company practices fractional assignment, I have some good news. You can instantly improve productivity by reassigning people to only one project at a time. Fractional assignment is dreadfully counterproductive: fractional workers don't bond with their teams; they often aren't around to hear conversations and answer questions and they must task switch, which incurs a significant hidden penalty. "The minimum penalty is 15 percent... Fragmented knowledge workers may look busy, but a lot of their busyness is just thrashing." [DeMarco 2002] (p.19-20) That's not to say that everyone needs to work with the team for the entire duration of the project. You can bring someone in to consult on a problem temporarily. However, while she works with the team, she should be fully engaged and available. http://www.jamesshore.com/Agile-Book/the_xp_team.html Tom DeMarco and Barry Boehm. 2002. The Agile Methods Fray. *Computer* 35, 6 (June 2002)
Frequent Verification And Validation	The terms Verification and Validation are commonly used in software engineering to mean two different types of analysis. The usual definitions are: **Validation**: Are we building the right system? **Verification**: Are we building the system right? In other words, validation is concerned with checking that the system will meet the customer's

	actual needs, while verification is concerned with whether the system is well-engineered, error-free, and so on. Verification will help to determine whether the software is of high quality, but it will not ensure that the system is useful. The distinction between the two terms is largely to do with the role of specifications. Validation is the process of checking whether the specification captures the customer's needs, while verification is the process of checking that the software meets the specification. Verification includes all the activities associated with the producing high quality software: testing, inspection, design analysis, specification analysis, and so on. It is a relatively objective process, in that if the various products and documents are expressed precisely enough, no subjective judgements should be needed in order to verify software. http://www.easterbrook.ca/steve/2010/11/the-difference-between-verification-and-validation/
Ideal Time	Like Work Units, Ideal Time excludes non-programming time. When a team uses Ideal Time for estimating, they are referring explicitly to only the programmer time required to get a feature or task done, compared to other features or tasks. Again, during the first few iterations, estimate history accumulates, a real velocity emerges, and Ideal Time can be mapped to real, elapsed time. Many teams using Ideal Time have found that their ultimate effort exceeds initial programmer estimates by 1-2x, and that this stabilizes, within an acceptable range, over a few iterations. On a task by task basis the ratio will vary, but over an entire iteration, the ratios that teams develop have proven to remain pretty consistent. For a given team, a known historical ratio of Ideal Time to real time can be especially valuable in planning releases. A team may quickly look at the required functionality and provide a high level estimate of 200 ideal days. If the team's historical ratio of ideal to real is about 2.5, the team may feel fairly confident in submitting an estimate of 500 project days. In fixed-bid scenarios, this kind of estimate can be reliable. https://www.versionone.com/agile-101/agile-management-practices/agile-feature-estimation/
Information Radiator	An information radiator is a large display of critical team information that is continuously updated and located in a spot where the team can see it constantly. The term "information radiator" was introduced extensively with a solid theoretical framework in Agile Software Development by Alistair Cockburn. Information radiators are typically used to display the state of work packages, the condition of tests or the progress of the team. Team members are usually free to update the information radiator. Some information radiators may have rules about how they are updated. Whiteboards, flip charts, poster boards or large electronic displays can all be used as the base media for an information radiator. For teams new to adopting agile work practices the best medium is usually a poster board on the wall with index cards and push pins. The index cards have a small amount of information on each of them and the push pins allow them to be moved around. Information radiators help amplify feedback, empower teams and focus a team on work results. Too many information radiators become confusing to understand and cumbersome to maintain. If an information radiator is not being updated it should be reconsidered and either changed or discarded. Robert McGeachy recommended a list of information radiators to be a part of every team room. Apart from the standard Scrum Artifacts, his list included,

	• Team structure with who's on the team • Client Organization structure • High Level and Mid Level plans • Client Phase exit criteria • Team performance survey results • Risks • Recognition awards • Ground Rules http://www.agileadvice.com/archives/2005/05/information_rad.html
Internal Rate Of Return (IRR)	Internal rates of return are commonly used to evaluate the desirability of investments or projects. The higher a project's internal rate of return, the more desirable it is to undertake the project. Assuming all projects require the same amount of up-front investment, the project with the highest IRR would be considered the best and undertaken first. A firm (or individual) should, in theory, undertake all projects or investments available with IRRs that exceed the cost of capital. Investment may be limited by availability of funds to the firm and/or by the firm's capacity or ability to manage numerous projects. In more specific terms, the IRR of an investment is the discount rate at which the net present value of costs (negative cash flows) of the investment equals the net present value of the benefits (positive cash flows) of the investment. http://en.wikipedia.org/wiki/Internal_rate_of_return
Iteration And Release Planning	Release planning is the process of transforming a product vision into a product backlog. The release plan is the visible and estimated product backlog itself, overlaid with the measured velocity of the delivery organization; it provides visual controls and a roadmap with predictable release points. http://www.netobjectives.com/files/Lean-AgileReleasePlanning.pdf
I.N.V.E.S.T.	Attributes of an effective user story: <table><tr><td>Independent</td><td>The user story should be self-contained, in a way that there is no inherent dependency on another user story.</td></tr><tr><td>Negotiable</td><td>User stories, up until they are part of a Sprint, can always be changed and rewritten.</td></tr><tr><td>Valuable</td><td>A user story must deliver value to the end user.</td></tr><tr><td>Estimable</td><td>You must always be able to estimate the size of a user story.</td></tr><tr><td>Sized appropriately or Small</td><td>User stories should not be so big as to become impossible to plan/task/prioritize with a certain level of certainty.</td></tr><tr><td>Testable</td><td>The user story or its related description must provide the necessary information to make test development</td></tr></table> http://www.agileforall.com/2009/05/new-to-agile-invest-in-good-user-stories/
Kaizen	Kaizen is the practice of continuous improvement. Kaizen was originally introduced to the West by Masaaki Imai in his book Kaizen: The Key to Japan's Competitive Success in 1986. Today Kaizen is recognized worldwide as an important pillar of an organization's long-term competitive

	strategy. http://www.kaizen.com/about-us/definition-of-kaizen.html
Kanban Boards	A Kanban means a ticket describing a task to do. A Kanban Board shows the current status of all the tasks to be done within this iteration. The tasks are represented by cards (Post-It Notes), and the statuses are presented by areas on the board separated and labeled with the status. This Kanban Board helps the team understand how they are doing well as well as what to do next and makes the team self-directing. http://www.infoq.com/articles/agile-kanban-boards http://blog.brodzinski.com/2009/11/kanban-story-kanban-board.html
Kano Model	The Kano model offers some insight into the product attributes which are perceived to be important to customers. The purpose of the tool is to support product specification and discussion through better development team understanding. Kano's model focuses on differentiating product features, as opposed to focusing initially on customer needs. Kano also produced a methodology for mapping consumer responses to questionnaires onto his model. The model involves two dimensions: • Achievement (the horizontal axis) which runs from *the supplier didn't do it at all* to *the supplier did it very well*. • Satisfaction (the vertical axis) that goes from *total dissatisfaction* with the product or service to *total satisfaction* with the product or service. Dr. Noriaki Kano isolated and identified three levels of customer expectations: that is, what it takes to positively impact customer satisfaction. The figure below portrays the three levels of need: expected, normal, and exciting. **Expected Needs** Fully satisfying the customer at this level simply gets a supplier into the market. The entry level expectations are the *must* level qualities, properties, or attributes. These expectations are also known as the *dissatisfiers* because by themselves they cannot fully satisfy a customer. However, failure to provide these basic expectations will cause dissatisfaction. Examples include attributes relative to safety, latest generation automotive components such as a self-starter, and the use of all new parts if a product is offered for sale as previously unused or new. The *musts* include customer assumptions, expected qualities, expected functions, and other *unspoken* expectations. **Normal Needs** These are the qualities, attributes, and characteristics that keep a supplier in the market. These next higher level expectations are known as the *wants* or the *satisfiers* because they are the ones that customers will specify as though from a list. They can either satisfy or

dissatisfy the customer depending on their presence or absence. The *wants* include *voice of the customer* requirements and other *spoken* expectations (see table below).

Exciting Needs The highest level of customer expectations, as described by Kano, is termed the *wow* level qualities, properties, or attributes. These expectations are also known as the *delighters* or *exciters* because they go well beyond anything the customer might imagine and ask for. Their absence does nothing to hurt a possible sale, but their presence improves the likelihood of purchase. *Wows* not only excite customers to make on-the-spot purchases but make them return for future purchases. These are *unspoken* ways of knocking the customer's socks off. Examples include heads-up display in a front windshield, forward- and rear-facing radars, and a 100,000 mile warranty. Over time, as demonstrated by the arrow going from top left to bottom right in the Kano model, *wows* become *wants* become *musts*, as in, for example, automobile self-starters and automatic transmissions.

http://asq.org/learn-about-quality/qfd-quality-function-deployment/overview/kano-model.html
http://en.wikipedia.org/wiki/Kano_model

KPIs in Agile	
	1. **Actual Stories Completed vs. Committed Stories** – the team's ability to understand and predict its capabilities. To measure, compare the number of stories committed to in sprint planning with the stories identified as completed in the sprint review.
	2. **Technical Debt Management** – the known problems and issues delivered at the end of the sprint. It is usually measured by the number of bugs, but may also include deliverables such as training material, user documentation and delivery media.
	3. **Team Velocity** – the consistency of the team's estimates from sprint to sprint. Calculate by comparing story points completed in the current sprint with points completed in the previous sprint; aim for +/- 10 percent.
	4. **Quality Delivered to Customers** – Are we building the product the customer needs? Does every sprint provide value to customers and become a potentially releasable piece of the product? It's not necessarily a product ready to release but rather a work in progress, designed to solicit customer comments, opinions and suggestions. This can best be measured by surveying the customers and stakeholders.
	5. **Team Enthusiasm** – a major component for a successful scrum team. If teammates aren't enthusiastic, no process or methodology will help. Measuring enthusiasm can be done by observing various sprint meetings or, the most straightforward approach, simply asking team members "Do you feel happy?" and "How motivated do you feel?"
	6. **Retrospective Process Improvement** – the scrum team's ability to revise its development process to make it more effective and enjoyable for the next sprint. This can be measured using the count of retrospective items identified, the retrospective items the team committed to addressing and the items resolved by the end of the sprint.
	7. **Communication** – how well the team, product owner, scrum master, customers and stakeholders are conducting open and honest communications. Through observing and listening you will get indications and clues about how well everyone is communicating.

	8. **Team's Adherence to Scrum Rules and Engineering Practices** – Although scrum doesn't prescribe engineering practices—unlike XP—most companies define several of their own for their projects. You want to ensure that the scrum team follows the rules your company defines. This can be measured by counting the infractions that occur during each sprint. 9. **Team's Understanding of Sprint Scope and Goal** – a subjective measure of how well the customer, product team and development team understand and focus on the sprint stories and goal. The goal is usually aligned with the intended customer value to be delivered and is defined in the acceptance criteria of the stories. This is best determined through day-to-day contact and interaction with the team and customer feedback. http://pragmaticmarketing.com/resources/9-scrum-metrics-to-keep-your-team-on-track
Lead Time	Lead time is a term borrowed from the manufacturing method known as Lean or Toyota Production System, where it is defined as the time elapsed between a customer placing an order and receiving the product ordered. http://guide.agilealliance.org/guide/leadtime.html

Lean Development	Lean software development is a translation of lean manufacturing and lean IT principles and practices to the software development domain.
	1. **Eliminate Waste**
	  • Provide market and technical leadership - your company can be successful by producing innovative and technologically advanced products but you must understand what your customers value and you know what technology you're using can deliver
	  • Create nothing but value - you have to be careful with all the processes you follow i.e. be sure that all of them are required and they are focused on creating value
	  • Write less code - the more code you have the more tests you need thus it requires more work and if you're writing tests for features that are not needed you are simply wasting time
	2. **Create Knowledge**
	  • Create design-build teams - leader of the development team has to listen to his/her members and ask smart questions encouraging them to look for the answers and to get back with encountered problems or invented solutions as soon as possible
	  • Maintain a culture of constant improvement - create environment in which people will be constantly improving what they are working on - they should know that they are not and should not be perfect - they always have a field to improve and they should do it
	  • Teach problem-solving methods - development team should behave like small research institute, they should establish hypotheses and conduct many rapid experiments in order to verify them
	3. **Build Quality In**
	  • Synchronize - in order to achieve high quality in your software you should start worrying about it before you write single line of working code - don't wait with synchronization because it will hurt
	  • Automate - automate testing, building, installations, anything that is routine, but do it smartly, do it in a way people can improve the process and change anything they want without worrying that after the change is done the software will stop working
	  • Refactor - eliminate code duplication to ZERO - every time it shows up refactor the code, the tests, and the documentation to minimize the complexity
	4. **Defer Commitment**

- Schedule Irreversible Decisions at the Last Responsible Moment - you should know where you want to go but you don't know the road very well, you will be discovering it day after day - the most important thing is to keep the right direction
- Break Dependencies - components should be coupled as loosely as possible to enable implementation in any order
- Maintain Options - develop multiple solutions for all critical decisions and see which one works best

5. **Optimize the Whole**
   - Focus on the Entire Value Stream - focus on winning the whole race which is the software - don't optimize local inefficiencies, see the whole and optimize the whole organization
   - Deliver a Complete Product - teams need to have great leaders as well as great engineers, sales, marketing specialists, secretaries, etc. - they together can deliver great final products to their customers

6. **Deliver Fast**
   - Work in small batches - reduce projects size, shorten release cycles, stabilize work environment (listen to what your velocity tells you), repeat what's good and eradicate practices that creates obstacles
   - Limit work to capacity - limit tasks queue to minimum (one or two iterations ahead is enough), don't be afraid of removing items from the queue - reject any work until you have an empty slot in your queue
   - Focus on cycle time, not utilization - put in your queue small tasks that cannot clog the process for a long time - reduce cycle time and have fewer things to process in your queue

7. **Respect People**
   - Train team leaders/supervisors - give team leaders the training, the guidance and some free space to implement lean thinking in their environment
   - Move responsibility and decision making to the lowest possible level - let your people think and decide on their own - they know better how to implement difficult algorithms and apply state-of-the-art software frameworks
   - Foster pride in workmanship - encourage passionate involvement of your team members to what and how they do

(EXAM TIP: Lots of questions about Lean, XP and Scrum principles)

http://www.disciplinedagiledelivery.com/lean-principles/

Learning Cycles	The Build–Measure–Learn loop emphasizes speed as a critical ingredient to product development. A team or company's effectiveness is determined by its ability to ideate, quickly build a minimum viable product of that idea, measure its effectiveness in the market, and learn from that experiment. In other words, it's a learning cycle of turning ideas into products, measuring customers' reactions and behaviors against built products, and then deciding whether to persevere or pivot the idea; this process repeats as many times as necessary. The phases of the loop are: Ideas –> Build –> Product –> Measure –> Data –> Learn https://en.wikipedia.org/wiki/Lean_startup
Minimally Marketable Feature (MMF)	Start by identifying your product's most desirable features. Prioritize their value, either through a formal technique or by subjectively ranking each one against the others. Once you've done so, plan your releases around the features. Release the highest-value features first to maximize their return. To accelerate delivery, have your entire team collaborate on one feature at a time and perform releases as often as possible.

	http://jamesshore.com/Articles/Business/Software%20Profitability%20Newsletter/Phased%20Releases.html
Minimal Viable Product (MVP)	A Minimum Viable Product is that version of a new product which allows a team to collect the maximum amount of validated learning about customers with the least effort. http://leanstack.com/minimum-viable-product/
MoSCoW	The MoSCoW approach to prioritization originated from the DSDM methodology (Dynamic Software Development Method), which was possibly the first agile methodology (?) – even before we knew iterative development as 'agile'. MoSCoW is a fairly simple way to sort features (or user stories) into priority order – a way to help teams quickly understand the customer's view of what is essential for launch and what is not. MoSCoW stands for: • **M**ust have (or Minimum Usable Subset) • **S**hould have • **C**ould have • **W**on't have (but Would like in future) http://www.allaboutagile.com/prioritization-using-moscow/
Multistage Integration Builds	Multi-stage Continuous Integration (MSCI) is an extension of the common practice of shielding others from additional changes by only checking-in when individual changes have been tested and only updating an individual workspace when it's ready to absorb other people's changes. With MSCI, each team does a team-based continuous integration first and then cross-integrates the team's changes with the mainline on success. This limits project-wide churn and allows continuous integration to scale to large projects. https://en.wikipedia.org/wiki/Multi-stage_continuous_integration
Net Present Value (NPV)	The difference between the present value of cash inflows and the present value of cash outflows. NPV is used in capital budgeting to analyze the profitability of an investment or project. NPV compares the value of a dollar today to the value of that same dollar in the future, taking inflation and returns into account. If the NPV of a prospective project is positive, it should be accepted. However, if NPV is negative, the project should probably be rejected because cash flows will also be negative. **Formula** Each cash inflow/outflow is <u>discounted</u> back to its present value (PV). Then they are summed. Therefore NPV is the sum of all terms, $$\frac{R_t}{(1+i)^t}$$ where t - the time of the cash flow i - the <u>discount rate</u> (the <u>rate of return</u> that could be earned on an investment in the financial markets with similar risk.); the opportunity cost of capital R_t- the net cash flow (the amount of cash, inflow minus outflow) at time t. For

	educational purposes, R_0 is commonly placed to the left of the sum to emphasize its role as (minus) the investment. http://en.wikipedia.org/wiki/Net_present_value

Negotiation	Negotiation, meaning "discussion intended to produce agreement", is fundamental to every software project. (And other projects too – my examples just happen to come from the software industry.) Developers and customers must reach agreement on what the system is supposed to do. A wise agreement will define achievable goals and meet the users' real needs.
	
	In Fisher and Ury's book, *Getting to Yes,* they call their approach "principled negotiation". It contains four key elements:
	  • Separate People from the Problem
	  • Focus on Interests, not Positions
	  • Invent Options for Mutual Gain
	  • Use Objective Criteria
	Both parties would be better served to engage in dialog about their underlying interests. Such dialog is encouraged by agile processes. They promote discussion, provide better opportunities to explore interests, and avoid premature "lock in" of positions. Fisher and Ury's style of negotiation is not about winning and losing. It's about *everybody winning*.
	
	http://www.agilekiwi.com/peopleskills/the-power-of-negotiation/
	Fisher, Roger and William Ury. Getting to Yes: Negotiating Agreement Without Giving In. New York, NY: Penguin Books, 1983.

Open Space Meetings	**Opening**: 1. Show the timeline (agenda), how the event breaks down into Opening, Marketplace of ideas, Break-out sessions, Closing. 2. Sponsor introduces the theme. Briefly. One or two minutes max. Long openings drain the energy of the meeting quickly. Get participants to work ASAP. 3. Facilitators introduce the principles and the format. Explain how the marketplace of ideas works. **Marketplace of ideas**: 1. Participants write 'issues' on pieces of paper. Preferably with bold markers, so they are easy to read from a distance. 2. Participants choose a timeslot for their topic on the agenda wall. 3. One by one, participants explain their issue to the others, with the aim of drawing the right people to their break-out-session. http://www.chriscorrigan.com/parkinglot/planning-an-open-space-technology-meeting/
Osmotic Communications	Osmotic communication means that information flows into the background hearing of members of the team, so that they pick up relevant information as though by osmosis. This is normally accomplished by seating them in the same room. Then, when one person asks a question, others in the room can either tune in or tune out, contributing to the discussion or continuing with their work. When osmotic communication is in place, questions and answers flow naturally and with surprisingly little disturbance among the team. http://alistair.cockburn.us/Osmotic+communication
Pareto Principle	Pareto's law is more commonly known as the 80/20 rule. The theory is about the law of

	distribution and how many things have a similar distribution curve. This means that *typically* 80% of your results may actually come from only 20% of your efforts! Pareto's law can be seen in many situations – not literally 80/20 but certainly the principle that the majority of your results will often come from the minority of your efforts. http://www.allaboutagile.com/agile-principle-8-enough-is-enough/
Parking Lot Charts	Parking Lot Charts summarize the top-level project status. A parking-lot chart contains a large rectangular box for each **theme** (or grouping of user stories) in a release. Each box is annotated with the name of the theme, the number of stories in that theme, the number of story points or ideal days for those stories, and the percentage of the story points that are complete. http://www.change-vision.com/en/visualizingagileprojects.pdf
Payback Period	The length of time required to recover the cost of an investment. http://www.investopedia.com/terms/p/paybackperiod.asp
Personas	A persona, first introduced by Alan Cooper, defines an archetypical user of a system, an example of the kind of person who would interact with it. The idea is that if you want to design effective software, then it needs to be designed for a specific person. For the bank, potential personas for a customer could be named Frances Miller and Ross Williams. In other words, personas represent fictitious people which are based on your knowledge of real users (EXAM TIP: Specific exam question on why use "extreme" personas) http://www.agilemodeling.com/artifacts/personas.htm
Planning Poker	The idea behind Planning Poker is simple. Individual stories are presented for estimation. After a period of discussion, each participant chooses from his own deck the numbered card that represents his estimate of how much work is involved in the story under discussion. All estimates are kept private until each participant has chosen a card. At that time, all estimates are revealed (the card is played) and discussion in differences between the estimates begins. The goal is to keep discussing until variance on the estimate only varies by 1. They highest final number is used for the product backlog. http://store.mountaingoatsoftware.com/pages/planning-poker-in-detail
Pre-mortem	Also called a Futurespective: The pre-mortem activity is great for preparing for an upcoming release or challenge. With a different perspective, the activity guides the participants to talk about all that could go wrong. Then the conversation switches to a mitigation and action plan. http://riskology.co/pre-mortem-technique/
Process Tailoring	You can decide to tailor up, or tailor down. This is up to each organization to decide based on its own business needs. You just need to state your approach in your guidelines and then follow it.. Always tailor up if you want to increase control on your project, simplify your planning/ tailoring process, and run projects with appropriate agility, efficiency, and discipline. http://www.enterpriseunifiedprocess.com/essays/softwareProcessImprovement.html
Product Roadmap	Product/Portfolio planning is a key activity for the Agile Product Manager, which usually consists

	of planning and management of existing product sets, and defining new products for the portfolio. Now, in order to define the portfolio, the product manager has to develop a **product roadmap** in collaboration with her stakeholders that consists of new upcoming products and existing product plan updates based on the their current status. The product roadmap thus enables identifying future release windows and drives planning for tactical development. http://www.agilejournal.com/articles/columns/column-articles/2650-product-road-mapping-using-agile-principles
Progressive Elaboration	Progressive elaboration is defined by The PMBOK® Guide as continuously improving and detailing a plan as more detailed and specific information and more accurate estimates become available as the project progresses, and thereby producing more accurate and complete plans that result from the successive iterations of the planning process. 1. Decide on a release timebox for the project. This may be one week, two weeks, one month. Whatever your team is comfortable with. 2. Look at the requirements on a high level and have the team decide approximately what you can release in each release cycle. Since this is a high level, approximate estimate, you don't need to be too detailed. It's just there to provide a rough idea about how the releases will develop. 3. At every iteration planning meeting, sit with the customer/product owner and decide what you are going to do in that iteration. At this stage, you can ask for more details, and the team can come up with a more accurate estimate based on the details that you now know. 4. At the end of the iteration, update the high level overview with any new information that you now have. 5. Repeat steps 3 and 4 for every iteration. https://projectmanagementessentials.wordpress.com/2010/04/07/progressive-elaboration-moving-from-the-unknown-to-the-known/
Refactoring	Refactoring (noun): a change made to the internal structure of software to make it easier to understand and cheaper to modify without changing its observable behavior. http://martinfowler.com/bliki/DefinitionOfRefactoring.html
Relative Prioritization and Ranking	You use planning, ranking, and priority to specify which work the team should complete first. If you rank user stories, tasks, bugs, and issues, all team members gain an understanding of the relative importance of the work that they must accomplish. Ranking and priority fields are used to build several reports. You rank and prioritize work items when you review the backlog for a product or iteration. http://www.processimpact.com/articles/prioritizing.html
Relative Sizing using Story Points	Your team collaboratively estimates each user story in story points. In his book "Agile Estimation and Planning," Mike Cohn defines story points this way: "Story points are a unit of measure for expressing the overall size of a user story, feature or other piece of work." Story points are relative values that do not translate directly into a specific number of hours. Instead, story points help a team quantify the general size of the user story. These relative estimates are less precise so that they require less effort to determine, and they hold up better over time. By estimating in story points, your team will provide the general size of the user stories now and develop the more detailed estimation of hours of work later, when team members are about to

	implement the user stories. https://www.excella.com/insights/sizing-agile-stories-with-the-relative-sizing-grid Mike Cohn. 2005. Agile Estimating and Planning. Prentice Hall PTR, Upper Saddle River, NJ, USA
Research Story	Sometimes programmers won't be able to estimate a story because they don't know enough about the technology required to implement the story. In this case, create a story to research that technology. An example of a research story is "Figure out how to estimate 'Send HTML' story". Programmers will often use a spike solution (see Spike Solutions) to research the technology, so these sorts of stories are often called spike stories. Programmers can usually estimate how long it will take to research a technology even if they don't know the technology in question. If they can't even estimate how long the research will take, timebox the story as you do with bug stories. I find that a day is plenty of time for most spike stories, and half a day is sufficient for most. http://jamesshore.com/Agile-Book/stories.html

Retrospectives	The meeting performed at the end of an iteration or Sprint in which the project team identifies opportunities for improvement in the next iteration, or in their Agile Process in its entirety.
	Approach
	1. Set the Stage
	   • Lay the groundwork for the session by reviewing the goal and agenda. Create an environment for participation by checking in and establishing working agreements.
	2. Gather Data
	   • Review objective and subjective information to create a shared picture. Bring in each person's perspective. When the group sees the iteration from many points of view, they'll have greater insight.
	3. Generate Insights
	   • Step back and look at the picture the team created. Use activities that help people think together to delve beneath the surface.
	4. Decide What to Do
	   • Prioritize the team's insights and choose a few improvements or experiments that will make a difference for the team.
	5. Close the Retrospective
	   • Summarize how the team will follow up on plans and commitments. Thank team members for their hard work. Conduct a little retrospective on the retrospective, so you can improve too.
	
	http://www.estherderby.com/tag/retrospectives
	Esther Derby and Diana Larsen. 2006. *Agile Retrospectives: Making Good Teams Great.* Pragmatic Bookshelf.

Return On Investment (ROI)	A performance measure used to evaluate the efficiency of an investment or to compare the efficiency of a number of different investments. To calculate ROI, the benefit (return) of an investment is divided by the cost of the investment; the result is expressed as a percentage or a ratio. If an investment does not have a positive ROI, or if there are other opportunities with a higher ROI, then the investment should be not be undertaken. The return on investment formula: $$ROI = \frac{(\text{Gain from Investment - Cost of Investment})}{\text{Cost of Investment}}$$ (EXAM TIP: The only finance question that I had was on one how the Product Manager manages

	ROI) http://www.investopedia.com/terms/r/returnoninvestment.asp
Risk Areas	Tom DeMarco and Tim Lister identified five core risk areas common to all projects in their book, Waltzing with Bears: • Intrinsic Schedule Flaw (estimates that are wrong and undoable from day one, often based on wishful thinking) • Specification Breakdown (failure to achieve stakeholder consensus on what to build) • Scope Creep (additional requirements that inflate the initially accepted set) • Personnel Loss • Productivity Variation (difference between assumed and actual performance) http://leadinganswers.typepad.com/leading_answers/2007/04/the_top_five_so.html Tom DeMarco and Timothy Lister. 2003. Waltzing with Bears: Managing Risk on Software Projects. Dorset House Publ. Co., Inc., New York, NY, USA
Risk Adjusted Backlog	Risk Adjusted Backlog focuses on where investment needs to be undertaken, based on risk. The normal risk assessment database process will provide a decreasing list of priorities from the risk calculation: Potential Consequence x Likelihood. It may be necessary to make decisions on which of these should be dealt with first within each of the risk bands. Ideally one would fund and rectify all high and significant risks within the current financial year. However, constraints on both funding and the time to prepare and complete work may cause this ideal to be delayed. https://refinem.com/essential-agile-processes-part-8-risk-adjusted-backlog/
Risk Based Spike	Spikes, another invention of XP are a special type of story used to drive out risk and uncertainty in a user story or other project facet. Spikes may be used for a number of reasons: 1. Spikes may be used for basic research to familiarize the team with a new technology of domain 2. The story may be too big to be estimated appropriately and the team may use a spike to analyze the implied behavior so they can split the story into estimable pieces. 3. The story may contain significant technical risk and the team may have to do some research or prototyping to gain confidence in a technological approach that will allow them to commit the user story to some future timebox. 4. The story may contain significant functional risk, in that although the intent of the story may be understood, it is not clear how the system needs to interact with the user to achieve the benefit implied. http://www.scaledagileframework.com/spikes/
Risk Burn Down Graphs	Risk Burndown graphs are very useful for seeing if the total project risk is increasing or decreasing over time. It allows stakeholders to see instantly if we are reducing project risk.

Risk Burndown Graph

https://www.mountaingoatsoftware.com/blog/managing-risk-on-agile-projects-with-the-risk-burndown-chart

Risk Exposure	Probability of a risk multiplied by the impact (in days) if the risk occurs. (The damage done by a risk occurring.) http://blog.mountaingoatsoftware.com/managing-risk-on-agile-projects-with-the-risk-burndown-chart
Risk Maps **Risk Heat Maps**	Whether you are managing a large program or a small project you will need to constantly highlight the key risks to your project or program steering group. One of the best ways I know to visually display risks in a succinct manner is to use a Risk Map (also known as a Risk Heat Map).On the vertical axis you have the probably of a given risk occurring, that is, the likelihood that the risk will materialize and become an Issue. On the horizontal axis we have the impact that the risk will have on the project or program should it materialize. One of the benefits of this method of displaying risks is that it's easy to see how risky the program or project is. If all the risks are clustered in the top right of the diagram then clearly your managing a very risky program or project http://www.expertprogrammanagement.com/2009/06/visualise-risks-using-a-risk-map/

Risk Multipliers	Risk multipliers account for common risks, such as turnover, changing requirements, work disruption, and so forth. These risk multipliers allow you to set a date, estimate how many story points of work you'll get done, and be right. It's a simpler version of the risk curves you'll see in good books on estimating and project management. http://jamesshore.com/Blog/Use-Risk-Management-to-Make-Solid-Commitments.html
Scrum Ceremonies	There are four ceremonies • Sprint planning: the team meets with the product owner to choose a set of work to deliver during a sprint • Daily scrum: the team meets each day to share struggles and progress • Sprint reviews: the team demonstrates to the product owner what it has completed during the sprint • Sprint retrospectives: the team looks for ways to improve the product and the process (EXAM TIP: We did not go deeply into the Scrum Roles or Ceremonies in this Guide since this is the most commonly trained and well documented material. If you have not already attended a Scrum training, you should read Ken Schwaber's *Agile Project Management with Scrum*, or minimally, the latest Scrum guide: http://www.scrumguides.org
Servant Leadership	There are several differences between Traditional projects and true Agile projects that—from a project management perspective—can best be summed up by the concept of self organization. In traditional projects, the project manager not only provides the vision of the team, but also directs and manages the team on the more detailed daily tasks by maintaining an up to date project plan. This usually results in a leadership style perhaps best described as "command and control." Agile projects on the other hand, still include the concepts of planning, managing the work, and providing status, but these activities are addressed collectively by the team, because at the end of the day they are the ones most familiar with what actually needs to happen to accomplish the project's goals. In this case, the Project Leader is in more of a support and facilitation role, similar in concept to Robert Greenleaf's idea of the Servant Leader. As Mike Cohn puts it in his Certified Scrum Master Class, the Project Leader's primary responsibilities are to "move boulders and carry water"—in other words, remove obstacles that prevent the team from providing business value, and make sure the team has the environment they need to succeed. One model often used to describe the leadership style needed on agile projects is the Collaborative Leadership model suggested by Pollyanna Pixton: • Make sure you have the **right people** on the project team. The right people are defined as those individuals who have passion about the goal of the project, have the ability to do the project, and are provided with the proper capacity, or time to work on the project. • **Trust First**, rather than waiting for people to prove their trustworthiness. • **Let the team members propose the approach** to make the project a success. After all, they are the ones who best know how to do the work. • **Stand back** and let the team members do their work without hovering over them continuously asking for status or trying to direct their activities, and provide support along the way to make sure nothing gets in the way of their success http://www.projectconnections.com/articles/092806-mcdonald.html Pollyanna Pixton, Niel Nickolaisen, Todd Little, and Kent McDonald. 2009. Stand Back and Deliver: Accelerating Business Agility (1st ed.). Addison-Wesley Professional
Shu Ha Ri	Shu-Ha-Ri is a way of thinking about how you learn a technique. The name comes from Japanese martial arts (particularly Aikido), and Alistair Cockburn introduced it as a way of thinking about learning techniques and methodologies for software development.

	The idea is that a person passes through three stages of gaining knowledge: • **Shu:** In this beginning stage the student follows the teachings of one master precisely. He concentrates on how to do the task, without worrying too much about the underlying theory. If there are multiple variations on how to do the task, he concentrates on just the one way his master teaches him. • **Ha:** At this point the student begins to branch out. With the basic practices working he now starts to learn the underlying principles and theory behind the technique. He also starts learning from other masters and integrates that learning into his practice. • **Ri:** Now the student isn't learning from other people, but from his own practice. He creates his own approaches and adapts what he's learned to his own particular circumstances. • http://martinfowler.com/bliki/ShuHaRi.html
Signal Card	English translation of the Japanese word, Kanban
Spike Solutions	A spike solution, or spike, is a technical investigation. It's a small experiment to research the answer to a problem. For example, a programmer might not know whether Java throws an exception on arithmetic overflow. A quick ten-minute spike will answer the question. http://jamesshore.com/Agile-Book/spike_solutions.html
Sprint Review	At the end of each sprint a sprint review meeting is held. During this meeting the Scrum team shows what they accomplished during the sprint. Typically this takes the form of a demo of the new features. The sprint review meeting is intentionally kept very informal, typically with rules forbidding the use of PowerPoint slides and allowing no more than two hours of preparation time for the meeting. A sprint review meeting should not become a distraction or significant detour for the team; rather, it should be a natural result of the sprint. Participants in the sprint review typically include the Product Owner, the Scrum team, the ScrumMaster, management, customers, and developers from other projects. During the sprint review the project is assessed against the sprint goal determined during the Sprint planning meeting. Ideally the team has completed each product backlog item brought into the sprint, but it is more important that they achieve the overall goal of the Sprint. http://www.mountaingoatsoftware.com/scrum/sprint-review-meeting
Story Maps	User story mapping offers an alternative for traditional agile planning approaches like the Scrum product backlog. Instead of a simple list, stories are laid out as a two dimensional map. The map provides both a high level overview of the system under development and of the value it adds to the users (the horizontal axis), and a way to organize detailed stories into releases according to importance, priority, etc. (the vertical axis). The map shows how every user story fits in the full scope. Releases are defined by creating horizontal slices of user stories, each slice is a release. For the first release, it is recommended to build a minimal set of user stories covering all user goals, so that you build a minimal but complete system to validate functionality and architecture early. https://www.thoughtworks.com/insights/blog/story-mapping-visual-way-building-product-backlog
Sustainable Pace	To set your pace you need to take your iteration ends seriously. You want the most completed, tested, integrated, production ready software you can get each iteration. Incomplete or buggy software represents an unknown amount of future effort, so you can't measure it. If it looks like

	you will not be able to get everything finished by iteration end have an iteration planning meeting and re-scope the iteration to maximize your project velocity. Even if there is only one day left in the iteration it is better to get the entire team re-focused on a single completed task than many incomplete ones. Working overtime sucks the spirit and motivation out of your team. When your team becomes tired and demoralized they will get less work done, not more, no matter how many hours are worked. Becoming over worked today steals development progress from the future. You can't make realistic plans when your team does more work this month and less next month. Instead of pushing people to do more than humanly possible use a release planning meeting to change the project scope or timing. Fred Brooks made it clear that adding more people is also a bad idea when a project is already late. The contribution made by many new people is usually negative. Instead ramp up your development team slowly well in advance, as soon as you predict a release will be too late. A sustainable pace helps you plan your releases and iterations and keeps you from getting into a death march. Find your team's perfect velocity that will remain consistent for the entire project. Every team is different. Demanding this team increase velocity to match that team will actually lower their velocity long term. So whatever your team's velocity is just accept it, guard it, and use it to make realistic plans. http://www.extremeprogramming.org/rules/overtime.html
Tail Length	The tail is the time period from "code slush" (true code freezes are rare) or "feature freeze" to actual deployment. This is the time period when companies do some or all of the following: beta testing, regression testing, product integration, integration testing, documentation, defect fixing. http://www.allaboutagile.com/shortening-the-tail/
Task Boards	Similar to Kanban Boards, a task board tracks the progress of work that is part of an overall story. https://realtimeboard.com/blog/scrum-kanban-boards-differences/
Team Development Stages	**Tuckman's Group Development Model** **Forming** In the *first stages* of team building, the *forming* of the team takes place. The individual's behavior is driven by a desire to be accepted by the others, and avoid controversy or conflict. Serious issues and feelings are avoided, and people focus on being busy with routines, such as team organization, who does what, when to meet, etc. But individuals are also gathering information and impressions - about each other, and about the scope of the task and how to approach it. This is a comfortable stage to be in, but the avoidance of conflict and threat means that not much actually gets done. The team meets and learns about the opportunities and challenges, and then agrees on goals and begins to tackle the tasks. Team members tend to behave quite independently. They may be motivated but are usually relatively uninformed of the issues and objectives of the team. Team members are usually on their best behavior but very focused on themselves. Mature team members begin to model appropriate behavior even at this early phase. Sharing the knowledge of the concept of "Teams - Forming, Storming, Norming, Performing" is extremely helpful to the team. Supervisors of the team tend to need to be directive during this phase. The forming stage of any team is important because, in this stage, the members of the team get to know one another, exchange some personal information, and make new friends. This is also a good opportunity to see how each member of the team works as an individual and how they

	respond to pressure. **Storming** Every group will next enter the *storming* stage in which different ideas compete for consideration. The team addresses issues such as what problems they are really supposed to solve, how they will function independently and together and what leadership model they will accept. Team members open up to each other and confront each other's ideas and perspectives. In some cases *storming* can be resolved quickly. In others, the team never leaves this stage. The maturity of some team members usually determines whether the team will ever move out of this stage. Some team members will focus on minutiae to evade real issues. The *storming* stage is necessary to the growth of the team. It can be contentious, unpleasant and even painful to members of the team who are averse to conflict. Tolerance of each team member and their differences should be emphasized. Without tolerance and patience the team will fail. This phase can become destructive to the team and will lower motivation if allowed to get out of control. Some teams will never develop past this stage. Supervisors of the team during this phase may be more accessible, but tend to remain directive in their guidance of decision-making and professional behavior. The team members will therefore resolve their differences and members will be able to participate with one another more comfortably. The ideal is that they will not feel that they are being judged, and will therefore share their opinions and views. **Norming** The team manages to have one goal and come to a mutual plan for the team at this stage. Some may have to give up their own ideas and agree with others in order to make the team function. In this stage, all team members take the responsibility and have the ambition to work for the success of the team's goals. **Performing** It is possible for some teams to reach the *performing* stage. These high-performing teams are able to function as a unit as they find ways to get the job done smoothly and effectively without inappropriate conflict or the need for external supervision. By this time, they are motivated and knowledgeable. The team members are now competent, autonomous and able to handle the decision-making process without supervision. Dissent is expected and allowed as long as it is channeled through means acceptable to the team. Supervisors of the team during this phase are almost always participative. The team will make most of the necessary decisions. Even the most high-performing teams will revert to earlier stages in certain circumstances. Many long-standing teams go through these cycles many times as they react to changing circumstances. For example, a change in leadership may cause the team to revert to *storming* as the new people challenge the existing norms and dynamics of the team. http://en.wikipedia.org/wiki/Tuckman's_stages_of_group_development
Team Space	William Pietri put together a list of rules for great development spaces. Amongst the well documented suggestions like putting the team together, room for daily standup, enough whiteboards and information radiators other suggestions included, • Get collaboration-friendly desks – William suggested this as one of the big pitfalls. He mentioned that many companies would like to foster collaboration but end up having furniture which is hostile to it. • Minimize distractions – The recommended rules to minimize distractions for the development stations include no phones, no email or IM, no off-topic conversation, less foot traffic and executives stay on mute. • Only direct contributors sit in the room – No chickens and certainly not the receptionist nor the sales people who would mostly be on the phone.

	• Pleasant space – Good lighting, decent air, plants, decorations and snacks. http://www.infoq.com/news/2010/02/agile-team-spaces
Technical Debt Management	It's all "those *internal* things that you choose not to do now, but which will impede future development if left undone" [Ward Cunningham]. On the surface the application looks to be of high quality and in good condition, but these problems are hidden underneath. QA may even tell you that the application has quality and few defects, but there is still debt. If this debt isn't managed and reduced, the cost of writing/maintaining the code will eventually outweigh its value to customers. In addition, it has a real financial cost: The time developers spend dealing with the technical debt and the resulting problems takes away from the time they can spend doing work that's valuable to the organization. The hard-to-read code that underlies technical debt also makes it more difficult to find bugs. Again, the time lost trying to understand the code is time lost from doing something more valuable. http://www.infoq.com/articles/technical-debt-levison **Managing technical debt** • **Starting captured debt.** Even if it is just by encouraging developers to note issues as they are writing code in the comments of that code, or putting in place more formal peer review processes where debt is captured it is important to document debt as it accumulates. • **Start measuring debt.** Once captured, placing a value / cost to the debt created enables objective discussions to be made. It also enables reporting to provide the organization with transparency of their growing debt. I believe that this approach would enable application and product end of life discussions to be made earlier and with more accuracy. • **Adopt standard architectures and open source models.** The more people that look at a piece of code the more likely debt will be reduced. The simple truth of many people using the same software makes it simpler and less prone to debt. http://theagileexecutive.com/2010/09/01/forrester-on-managing-technical-debt/
Test First Development	Test-First programming involves producing automated unit tests for production code, before you write that production code. Instead of writing tests afterward (or, more typically, not ever writing those tests), you always begin with a unit test. For every small chunk of functionality in production code, you first build and run a small (ideally very small), focused test that specifies and validates what the code will do. This test might not even compile, at first, because not all of the classes and methods it requires may exist. Nevertheless, it functions as a kind of executable specification. You then get it to compile with minimal production code, so that you can run it and watch it fail. (Sometimes you expect it to fail, and it passes, which is useful information.) You then produce exactly as much code as will enable that test to pass. http://www.versionone.com/Agile101/Test-First_Programming.asp
Test-Driven Development	Test-Driven Development (TDD) is a special case of test-first programming that adds the element of continuous design. With TDD, the system design is not constrained by a paper design document. Instead you allow the process of writing tests and production code to steer the design as you go. Every few minutes, you refactor to simplify and clarify. If you can easily imagine a clearer, cleaner method, class, or entire object model, you refactor in that direction, protected the entire time by a solid suite of unit tests. The presumption behind TDD is that you

	cannot really tell what design will serve you best until you have your arms elbow-deep in the code. As you learn about what actually works and what does not, you are in the best possible position to apply those insights, while they are still fresh in your mind. And all of this activity is protected by your suites of automated unit tests. You might begin with a fair amount of up front design, though it is more typical to start with fairly modest design; some white-board UML sketches often suffice in the Extreme Programming world. But how much design you start with matters less, with TDD, than how much you allow that design to diverge from its starting point as you go. You might not make sweeping architectural changes, but you might refactor the object model to a large extent, if that seems like the wisest thing to do. Some shops have more political latitude to implement true TDD than others. http://www.versionone.com/Agile101/Test-First_Programming.asp
Throughput	Throughput is the amount of work items delivered in a given period of time (e.g. week, month, quarter). http://old.berriprocess.com/en/todas-las-categorias/item/62-analiticas-lean-kanban-rendimiento
Timeboxing	Timeboxing is a planning technique common in planning projects (typically for software development), where the schedule is divided into a number of separate time periods (timeboxes, normally two to six weeks long), with each part having its own deliverables, deadline and budget. Timeboxing is a core aspect of rapid application development (RAD) software development processes such as dynamic systems development method (DSDM) and agile software development. Scrum Timeboxes: • Sprint Planning – 2 sessions, 1 hour per week of sprint • Sprint Duration – 1-4 weeks • Daily Scrums – 15 mins/day • Sprint Review – 1 hours per week of sprint, 1 hour prep • Sprint Retrospective – 3 hours per sprint https://daymoframework.wordpress.com/2010/08/21/a-crash-course-in-time-boxing/
Trade-Off Matrix	Balancing the four constraints – compliance, cost, schedule, and scope – is not a trivial task. However, just like the Agile Triangle, the Tradeoff Matrix used in Agile software development applies to IT. In its software development variant, the Tradeoff matrix is an effective tool to decide between conflicting constraints, as follows:

	Fixed	Flexible	Accept
Scope			X
Schedule	X		
Cost		X	

Rules:

- *Fixed* trumps *Flexible* trumps *Accepts*
- Each column can contain only one check mark
- Two check marks can't have the same priority

Note: The specific check marks in Table 1 are merely illustrative. Any three check marks that adhere to the rules above are legitimate. In fact, the three check marks represent the organization's policy decision as to what really matters.

http://theagileexecutive.com/tag/tradeoff-matrix/

Transition Indicator	A transition indicator is a notification that a risk (i.e., something that will have a negative impact on the cost/schedule of the project *if* it occurs) has materialized and is in need of attention. http://www.informit.com/articles/article.aspx?p=2123314&seqNum=3
Usability Testing	Usability testing is a long-established, empirical and exploratory technique to answer questions such as "how would an end user respond to our software under realistic conditions?" It consists of observing a representative end user interacting with the product, given a goal to reach but no specific instructions for using the product. (For instance, a goal for usability testing of a furniture retailer's Web site might be "You've just moved and need to do something about your two boxes of books; use the site to find a solution.") http://guide.agilealliance.org/guide/usability.html
Use Cases	Use cases are sometimes used in heavyweight, control-oriented processes much like traditional requirements. The system is specified to a high level of completion via the use cases and then locked down with change control on the assumption that the use cases capture everything. Use cases attempt to bridge the problem of requirements not being tied to user interaction. A use case is written as a series of interactions between the user and the system, similar to a call and response where the focus is on how the user will use the system. In many ways, use cases are better than a traditional requirement because they emphasize user-oriented context. The value of the use case to the user can be divined, and tests based on the system response can be figured out based on the interactions. Use cases usually have two main components: Use case

	diagrams, which graphically describe actors and their use cases, and the text of the use case itself. http://en.wikipedia.org/wiki/Use_case
User Stories	A good way to think about a user story is that it is a reminder to have a conversation with your customer (in XP, project stakeholders are called customers), which is another way to say it's a reminder to do some just-in-time analysis. In short, user stories are very slim and high-level requirements artifacts. User stories are one of the primary development artifacts for Scrum and Extreme Programming (XP) project teams. A user story is a very high-level definition of a requirement, containing just enough information so that the developers can produce a reasonable estimate of the effort to implement it. http://www.agilemodeling.com/artifacts/userStory.htm
Value Stream Mapping	Value Stream Maps exist for two purposes: to help organizations identify and end wasteful activities. Finding problems and creating a more efficient process isn't easy; even the best organization can be made more efficient and effective. But bringing about substantive organizational change that actually eliminates waste is a tall order. It's comparatively easy to identify waste, but it's another matter entirely to stop waste from happening in the first place. Value Stream Maps can both sharpen an organization's skills in identifying waste and help drive needed change. But first things first: What a Value Stream Map is and how one can be intelligently produced. The examples and concepts that follow are based on applying a Value Stream Map to a software engineering organization, but these concepts are applicable to a wide range of settings. Value Stream Maps help us bring about organizational improvement, progress in our processes and methods, and most importantly, better software. Value Stream Maps can help both identify and stop waste in an organization http://www.ibm.com/developerworks/rational/library/10/howandwhytocreatevaluestreammapsforswengineerprojects/index.html?ca=drs-
Velocity	Velocity is an extremely simple, powerful method for accurately measuring the rate at which teams consistently deliver business value. To calculate velocity, simply add up the estimates of the features (user stories, requirements, backlog items, etc.) successfully delivered in an iteration. There are some simple guidelines for estimating initial velocity prior to completing the first iteration but after that point teams should use proven, historical measures for planning features. Within a short time, velocity typically stabilizes and provides a tremendous basis for improving the accuracy and reliability of both near-term and longer-term project planning. Agile delivery cycles are very small so velocity emerges and can be validated very early in a project and then relied upon to improve project predictability. http://www.versionone.com/agile-101/agile-scrum-velocity/
Wide Band Delphi	Estimation method is a consensus-based technique for estimating effort 1. Coordinator presents each expert with a specification and an estimation form. 2. Coordinator calls a group meeting in which the experts discuss estimation issues with the coordinator and each other. 3. Experts fill out forms anonymously. 4. Coordinator prepares and distributes a summary of the estimates 5. Coordinator calls a group meeting, specifically focusing on having the experts discuss

	points where their estimates vary widely 6. Experts fill out forms, again anonymously, and steps 4 to 6 are iterated for as many rounds as appropriate. http://en.wikipedia.org/wiki/Wideband_delphi
WIP Limits	Limiting work in process (WIP) to match your team's development capacity helps ensure the traffic density does not increase the capacity of your team. The Kanban board will help you get to the right WIP limit as you become better at it. Without WIP limits you will continue to pile up partially completed work in the pipe thereby creating the phantom traffic jam. Adding to your WIP without completing anything just increases the duration of all tasks in the queue. If you are a product development shop, having a large duration (lead time) can significantly affect your company's profitability. http://www.kanbanway.com/importance-of-kanban-work-in-progress-wip-limits
Wireframes	A wireframe is a "low fidelity" prototype. This non-graphical artifact shows the skeleton of a screen, representing its structure and basic layout. It contains and localizes contents, features, navigation tools and interactions available to the user. The wireframe is usually: • black and white, • accompanied by some annotations to describe the behavior of the elements (default or expected states, error cases, values, content source…), their relationships and their importance, • often put in context within a storyboard (a sequence of screens in a key scenario) • refined again and again • used as a communication tool serving as an element of conversation and confirmation of "agile" user stories http://www.agile-ux.com/tag/wireframe/
Agile Roles	
Scrum Roles	(EXAM TIP: I did not go deeply into the Scrum Roles or Ceremonies since this is the most commonly trained material. If you have not already attended a Scrum training, you should read Ken Schwaber's *Agile Project Management with Scrum,* or minimally, read the latest Scrum guide by him and Jeff Sutherland: http://www.scrumguides.org
Product Owner	The **product owner** decides what will be built and in which order. • Defines the features of the product or desired outcomes of the project • Chooses release date and content • Ensures profitability (ROI) • Prioritizes features/outcomes according to market value • Adjusts features/outcomes and priority as needed • Accepts or rejects work results • Facilitates scrum planning ceremony

	http://www.scrumguides.org
Scrum Master	The ScrumMaster is a facilitative team leader who ensures that the team adheres to its chosen process and removes blocking issues. • Ensures that the team is fully functional and productive • Enables close cooperation across all roles and functions • Removes barriers • Shields the team from external interferences • Ensures that the process is followed, including issuing invitations to daily scrums, sprint reviews, and sprint planning • Facilitates the daily scrums http://www.scrumguides.org
Team	• Is cross-functional • Is right-sized (the ideal size is seven -- plus/minus two -- members) • Selects the sprint goal and specifies work results • Has the right to do everything within the boundaries of the project guidelines to reach the sprint goal • Organizes itself and its work • Demos work results to the product owner and any other interested parties. http://www.scrumguides.org
Extreme Programming (XP) Roles	
XP Coach	The XP Coach role helps a team stay on process and helps the team to learn. A coach brings an outside perspective to help a team see themselves more clearly. The coach will help balance the needs of delivering the project while improving the use of the practices. A coach or team of coaches supports the Customer Team, the Developer Team, and the Organization. The decisions that coaches make should always stem from the XP values (communication, simplicity, feedback, and courage) and usually move toward the XP practices. As such, familiarity with the values and practices is a prerequisite. The coach must command the respect required to lead the respective teams. The coach must possess people skills and be effective in influencing the actions of the teams. http://epf.eclipse.org/wikis/xp/xp/roles/xp_coach_60023190.html
XP Customer	The XP Customer role has the responsibility of defining what is the right product to build, determining the order in which features will be built and making sure the product actually works. The XP Customer writes system features in the form of user stories that have business value. Using the planning game, he chooses the order in which the stories will be done by the development team. He also defines acceptance tests that will be run against the system to prove that the system is reliable and does what is required. The customer prioritizes user stories, the team estimates them. http://epf.eclipse.org/wikis/xp/xp/roles/xp_customer_6D7CB91B.html

XP Programmer	The XP Programmer is responsible for implementing the code to support the user stories http://epf.eclipse.org/wikis/xp/xp/roles/xp_programmer_D005E927.html-
XP Programmer (Administrator)	The XP Programmer (Administrator) role includes most of the traditional software development technical roles, such as designer, implementer, integrator, and administrator. In the administrator role, the programmer deals with establishing the physical working environment http://epf.eclipse.org/wikis/xp/xp/roles/xp_system_administrator_92735060.html
XP Tracker	The three basic things the XP Tracker will track are the release plan (user stories), the iteration plan (tasks) and the acceptance tests. The tracker can also keep track of other metrics, which may help in solving problems the team is having. A good XP Tracker has the ability to collect the information without disturbing the process significantly. http://epf.eclipse.org/wikis/xp/xp/roles/xp_tracker_AD8A6C9F.html
XP Tester	The primary responsibility of the XP Tester is to help the customer define and implement acceptance tests for user stories. The XP Tester is also responsible for running the tests frequently and posting the results for the whole team to see. As the number of tests grow, the XP Tester will likely need to create and maintain some kind of tool to make it easier to define them, run them, and gather the results quickly. Whereas knowledge of the applications target domain is provided by the customer, the XP Tester needs to support the customer by providing: <ul><li>Knowledge of typical software failure conditions and the test techniques that can be employed to uncover those errors.</li><li>Knowledge of different techniques to implement and run tests, including understanding of and experience with test automation</li></ul> http://epf.eclipse.org/wikis/xp/xp/roles/xp_tester_44877D41.html

The Agile Manifesto

In February 2001, 17 software developers met at the Snowbird, Utah resort, to discuss lightweight development methods. They published the *Manifesto for Agile Software Development* to define the approach now known as agile software development. Some of the manifesto's authors formed the Agile Alliance, a nonprofit organization that promotes software development according to the manifesto's principles.

The Agile Manifesto reads, in its entirety, as follows:

We are uncovering better ways of developing software by doing it and helping others do it. Through this work we have come to value:

Individuals and interactions over processes and tools

Working software over comprehensive documentation

Customer collaboration over contract negotiation

Responding to change over following a plan

That is, while there is value in the items on the right, we value the items on the left more.

The meanings of the manifesto items on the left within the agile software development context are described below:

- Individuals and Interactions – in agile development, self-organization and motivation are important, as are interactions like co-location and pair programming.
- Working software – working software will be more useful and welcome than just presenting documents to

clients in meetings.

- Customer collaboration – requirements cannot be fully collected at the beginning of the software development cycle, therefore continuous customer or stakeholder involvement is very important.
- Responding to change – agile development is focused on quick responses to change and continuous development.

Twelve principles underlie the Agile Manifesto, including:

1. Our highest priority is to satisfy the customer through early and continuous delivery of valuable software.
2. Welcome changing requirements, even late in development. Agile processes harness change for the customer's competitive advantage.
3. Deliver working software frequently, from a couple of weeks to a couple of months, with a preference to the shorter timescale.
4. Business people and developers must work together daily throughout the project.
5. Build projects around motivated individuals. Give them the environment and support they need, and trust them to get the job done.
6. The most efficient and effective method of conveying information to and within a development team is face-to-face conversation.
7. Working software is the primary measure of progress.
8. Agile processes promote sustainable development. The sponsors, developers, and users should be able to maintain a constant pace indefinitely.
9. Continuous attention to technical excellence and good design enhances agility.
10. Simplicity--the art of maximizing the amount of work not done--is essential.
11. The best architectures, requirements, and designs emerge from self-organizing teams.
12. At regular intervals, the team reflects on how to become more effective, then tunes and adjusts its behavior accordingly.

http://agilemanifesto.org/

www.ingramcontent.com/pod-product-compliance
Lightning Source LLC
Chambersburg PA
CBHW051211200326
41519CB00025B/7081